Advances in Oil and Gas Exploration & Production

Series Editor

Rudy Swennen, Department of Earth and Environmental Sciences,
K.U. Leuven, Heverlee, Belgium

The book series Advances in Oil and Gas Exploration & Production publishes scientific monographs on a broad range of topics concerning geophysical and geological research on conventional and unconventional oil and gas systems, and approaching those topics from both an exploration and a production standpoint. The series is intended to form a diverse library of reference works by describing the current state of research on selected themes, such as certain techniques used in the petroleum geoscience business or regional aspects. All books in the series are written and edited by leading experts actively engaged in the respective field.

The Advances in Oil and Gas Exploration & Production series includes both single and multi-authored books, as well as edited volumes. The Series Editor, Dr. Rudy Swennen (KU Leuven, Belgium), is currently accepting proposals and a proposal form can be obtained from our representative at Springer, Dr. Alexis Vizcaino (Alexis.Vizcaino@springer.com).

More information about this series at http://www.springer.com/series/15228

Troyee Dasgupta · Soumyajit Mukherjee

Sediment Compaction and Applications in Petroleum Geoscience

 Springer

Troyee Dasgupta
Department of Earth Sciences
Indian Institute of Technology Bombay
Mumbai, Maharashtra, India

Soumyajit Mukherjee
Department of Earth Sciences
Indian Institute of Technology Bombay
Mumbai, Maharashtra, India

ISSN 2509-372X   ISSN 2509-3738   (electronic)
Advances in Oil and Gas Exploration & Production
ISBN 978-3-030-13444-0   ISBN 978-3-030-13442-6   (eBook)
https://doi.org/10.1007/978-3-030-13442-6

Library of Congress Control Number: 2019935486

This Springer imprint is published by the registered company Springer Nature Switzerland AG
The registered company address is: Gewerbestrasse 11, 6330 Cham, Switzerland

Troyee Dasgupta *dedicates this book to her daughter "Rahini Dasgupta, born on 31-Jan-2017".*

Soumyajit Mukherjee *dedicates this book to Profs. Joydip Mukhopadhyay and Prabir Dasgupta for teaching sedimentology and stratigraphy in great detail during his B.Sc. studies in the then Presidency College (Kolkata) during 1996–1999.*

Acknowledgements

TDG wrote this book while in maternity leave (February to August 2017). SM co-authored while in research sabbatical in IIT Bombay in 2017. We acknowledge our spouses, Swagato Dasgupta and Payel Mukherjee, respectively, for their support. TDG is thankful to Dr. Lalaji Yadav for his fruitful mentorship in Petrophysics. Sandeep Gaikwad (IIT Bombay) drew few diagrams. Thanks to Helen Ranchner, Annett Beuttner and the proof-reading team (Springer).

Contents

Symbols

Φ	Porosity
Φ_0	Average surface porosity of the surface clays
c	A constant
z	Burial depth
ρ_h	Hydrostatic pressure
γ_w	Specific weight of water
h	Height of column of water
G_h	Hydrostatic pressure gradient
P	Pore pressure
σ_v	Overburden stress
σ_e	Effective stress
α	Biot's effective stress coefficient
Rn	Resistivity normal trend
R	Resistivity log
X	Normal compaction trend
Δt	Interval transit time
Δt_n	Interval transit time normal trend
Y	Pore pressure gradient
P_f	Formation fluid pressure
α_v	Normal overburden stress gradient
β	Normal fluid pressure gradient
Z	Depth
Δt	Sonic transit time
A, B	Parameters
P_B	Pore pressure
σ_A	Effective stress at A
P_{NA}	Hydrostatic normal pore pressure at point A
OB_B	Overburden pressure at point B
OB_A	Overburden pressure at point A
σ_M	Mean effective stress
σ	Vertical effective stress

σ_h	Minimum horizontal stress
σ_H	Maximum horizontal effective stress
V	Sonic velocity
V_{min}	Minimum sonic velocity of the rock matrix
V_{max}	Maximum sonic velocity of the rock matrix
Σ	Vertical effective stress
P	Pore pressure
ρ_{max}	Maximum matrix density
ρ_f	Fluid density
Δt_f	Interval transit time of fluid
Δt_n	Interval transit time for the normal pressure in shales
Δt	Transit time of shale
V_p	Compressional wave velocity
V_{ml}	Mudline velocity
U	Parameter representing uplift of the sediments
σ_{max}	Effective stress
v	Velocity
V_m	Sonic interval velocity with the shale matrix
a_m	Ratio of the loading and unloading velocities in the effective stress curves
V_{max}	Velocity at the start of unloading
P_{ulo}	Pore pressure due to unloading
$\emptyset_{RHOB}$	Porosity from density log
ρ_{ma}	Matrix density
ρ_b	Bulk density measured by log
ρ_{fl}	Fluid density
Δt_{ma}	Interval transit time of the matrix
Δt_{fl}	Fluid transit time
Δt	Average interval transit time from log
$\emptyset_{DT}$	Porosity from sonic log
$\emptyset_{RILD}$	Porosity from resistivity log
R_w	Formation water resistivity
n	Saturation exponent
m	Cementation exponent
Rt	True resistivity of the formation
t_{ma}	Sonic transit time of the rock matrix
ϕz	Porosity at depth z
ϕ_0	Porosity at the surface
b	A constant
Δt	Transit time measured by the sonic log

Δt_o Transit time at the present sedimentary surface

c   Compaction coefficient

z   Burial depth

Δt_o Transit time near to the transit time of water

List of Figures

List of Tables

Compaction of Sediments and Different Compaction Models

1

Abstract

Various simple and advanced models exist for mechanisms of uniform and non-uniform sediment compaction that increases density and reduces porosity. While the classical Athy's relation on depth-wise exponential reduction of porosity is not divided into any distinct stages, the Hedberg's model involves four stages. Weller's model utilized Athy's and Hedberg's relations to deduce a sediment compaction model. Power's compaction model additionally considers clay mineralogy. Several other porosity/compaction models exist, e.g., those by Teodorovich and Chernov, Burst, Beall, and Overton and Zanier. The geometry of the depth-wise porosity profile depends on the sedimentation rate, compaction mechanism and pressure solution model. This chapter reviews porosity variation with depth for the following rock types: shales, shaly sandstones, sandstones and carbonates.

1.1 Introduction

The chemical and the physical properties of sediments and sedimentary rocks alter as the overburden pressure increases. These changes relate to burial depth, temperature and time. Experiments by Warner (1964) suggested that acceleration of the rate of compaction of sediments seem to be the only change at <200 °F. Compaction of sediments reduces porosity and increases density (Bjørlykke et al. 2009). The reduction of porosity is a convenient way of measuring the amount of sediments compacted since deposition took place, for practical purposes. Empirical compaction curves are the plots of porosity versus depth up to ∼6 km. Mechanical compaction being the primary mechanism of compaction, clay minerals are often utilized in many models to visualise how grains rearrange with depth. The composition varies from proximal to distal part of the basin and the compaction pattern of each sediment type differs (Bjørlykke et al. 2009). Compaction models explain the major processes for the sediment compaction. This helps the interpreters to visualise the relationship of porosity loss with depth and the probable reason for anomalous zones. The evolution of compaction models and porosity reduction with depth from different parts of the world are presented in Fig. 1.1.

1.2 Porosity Models

Sediment porosities undergo changes with burial. In the consecutive sections the different models are explained.

© Springer Nature Switzerland AG 2020
T. Dasgupta and S. Mukherjee, *Sediment Compaction and Applications in Petroleum Geoscience*,
Advances in Oil and Gas Exploration & Production, https://doi.org/10.1007/978-3-030-13442-6_1

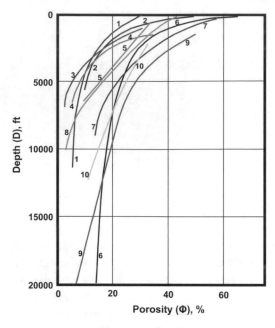

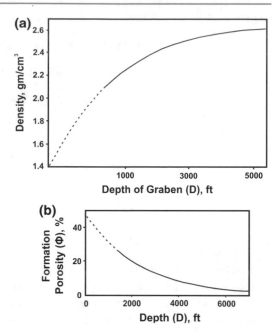

Fig. 1.1 Compaction models shown by porosity versus depth of burial. Inter-relationship for shales and argillaceous sediments: 1–9: Different curves. Modified after Fig. 17 of Rieke and Chilingarian (1974)

Fig. 1.2 **a** Bulk density versus depth relationship for shales from Oklahoma (U.S.A). **b** Porosity versus depth relationship for shales from Oklahoma (U.S.A) (Modified after Fig. 14 of Rieke and Chilingarian 1974)

1.2.1 Athy's Model

The compaction model by Athy (1930a) states that porosity falls with depth by expulsion of interstitial fluids. Pure shale samples from Oklahoma exhibit a definite relation between porosity and depth (Fig. 1.2). In the cycle of sediment deposition up to burial, several processes play an important role in porosity change viz. (i) deformation of grains and granulation; (ii) cementation; (iii) dissolution; (iv) recrystallisation; (v) squeezing of grains. Porosity reduction and density increase are directly proportional to the increase of overburden and tectonic stresses, but the degree of compaction is neither related to porosity loss nor to the increase in density (Athy 1930b). Athy's model has been widely referred in sedimentology text books (e.g., Pettijohn 1984), and in tectonic models (Mukherjee 2017, 2018a, b, c). Athy's model is certainly in marked contrast with other available models of sediment compaction where porosity is not at all considered (such as Mukherjee and Kumar 2018).

Athy's (1930a) widely applied algebraic relation demonstrate the exponential porosity decrease from 0.5 to 0.05 porosity unit (pu) from surface up to 2.3 km:

$$\Phi = \Phi_0 \cdot e^{-cz} \qquad (1.1)$$

Here Φ: porosity, Φ_0: average surface porosity of the surface clays, c: a constant, and z: burial depth.

1.2.2 Hedberg's Model

Hedberg (1936) established a compaction curve by determining porosities mainly from mudrocks from core samples of the Venezuelan wells. The depth of samples ranges 291–6175 ft (88.6–1882.14 m). Hedberg (1936) defined compaction in four stages. The first stage involves a mechanical rearrangement and dewatering of sediments including the adsorbed water. Porosity reduces from 90 to 75% with increase in overburden pressure. Loss of adsorbed water characterizes the next stage of porosity reduction from 75 to 35%. Below 35% porosity, the clay particles approach mechanical deformation and this resists further reduction in porosity. Below 35% porous clays alter to shales and this phenomenon increases the rigidity in the grain structure (Hamilton 1959). With large increments of pressure, porosity reduction is slow below 10% of porosity. In the third stage, grains recrystallize generating porosities <10%. The final fourth stage is dominated by chemical readjustment (Hunt et al. 1998).

The differences between Athy's and Hedberg's curves (Fig. 1.3) are due to factors such as temperature, different ages of the studied samples etc. These curves define the mudrock compaction. Athy's curve generally represents compaction of the mudrock.

1.2.3 Weller's Model

Weller (1959) used Athy's, Hedberg's and Terzaghi's data to define compaction curves. Weller's compaction process resembles that of

Hedberg. The reduction of porosity is from 85% at surface to 45%. Further burial promotes porosity reduction from 45 to 10%. Weller (1959) stated that the clay particles occupy the void spaces as the *"non-clay particles deform and share mutual contact"*.

1.2.4 Power's Model

Power's (1967) compaction model is based on changes of clay mineralogy with burial. During deep burial of sediments, large volume of water is expelled with subsequent transformation of clay composition from montmorillonite to illite. Power (1967) explained the clay transformation and the changes in the adsorbed water content at different depths. Power's (1967) theory is well explained in Fig. 1.4 and in its caption.

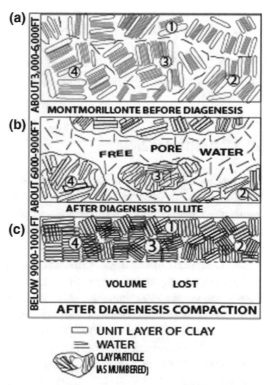

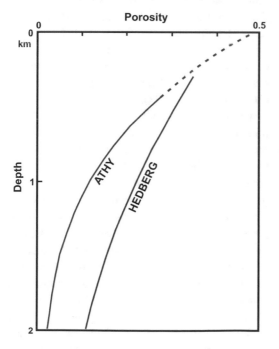

Fig. 1.3 Comparison between Athy's and Hedberg's compaction curves. Modified after Chapman (1994)

Fig. 1.4 Different stages of compaction effect on clay diagenesis. **a** Clays consist of bound water at the time of deposition. **b** Free water increases with burial with the consequent release of boundwater. **c** Free water is squeezed out and the original volume reduces (Fig. 57 in Rieke and Chilingarian 1974)

1.2.5 Teodorovich and Chernov's Model

Teodorovich and Chernov (1968) explained a three-stage compaction. The first stage resembles Athy's theory where the initial porosity loss is fast i.e., from 66 to 40% for clays and sandstones, and 56 to 40% for siltstones. Large volume of fluid release takes place in this stage. The burial depth ranges 0 to 8–10 m. In the second stage, porosity falls sharply from 8–10 m to 1200–1400 m. The porosity loss for shales, siltstones and sandstones are similar which is ~20%. Slow compaction prevails in the third/final stage at 1400–6000 m burial depth where the shale porosity ranges 7–8% and sandstones-siltstones ~15–16%. Teodorovich and Chernov (1968) theory utilized data from the Azerbaijan area.

1.2.6 Burst's Model

The Burst (1969) compaction model resembles the previous models of clay transformation and dehydration.

1.2.7 Beall's Model

This model is based on core data obtained from well samples from offshore Louisiana. The results are obtained from high pressure experiments on drilling muds from deep sea drilling programme of JOIDES. This model discusses variation of pore throat angle with subsequent sediment burial. Similar to the previous theories, Beall's model proposes the expulsion of fluids during initial stages of sediment burial with pore throat angle ~6 around 3300 ft (1006 m) depth. The clays approach an angle of 1 Å ~ at 3300–8000 ft (2438 m) where the sediments are 75% compacted. The porosity decreases at a slower rate during the third stage of compaction where the angle is >1 Å. Beall also explained the expulsion of less saline fluids during this stage whereby the dissolved NaCl enters the nearby permeable sand beds.

1.2.8 Overton and Zanier's Model

Overton and Zanier's (1970) model explains the compaction of sediments in four stages and resembles Beall's model. The model mentions the different water types in four stages. Fresh water is predominant at <3000 ft (914 m) depth. Salinity increases exponentially between 3000 and 10,000 ft (914.4–3048 m) depths. Depths exceeding 10,000 ft (3048 m) have highest pressure gradients where the salinity falls with increasing depth. At depths >15,000 ft (4572 m), salinity is more than the upper zone and the water percentage is lower in this zone.

1.3 Normal Porosity Profiles for Clastics

1.3.1 General Points

Worldwide there are different types of basins with various sediment deposition patterns (recent reviews in Bushby and Azor 2012; Allen and Allen 2014). Normally consolidated sediments of uniform lithology show normal porosity profiles (NPP) (Burrus 1998). Increasing vertical stress can also reduce porosity (Hunt et al. 1998). The normal porosity profiles can be reconstructed by calculating porosities from well logs, but this is not straightforward because of the constraints of each log (Burrus 1998).

Perrier and Quibler (1974) introduced the normal porosity profile (NPP) to calculate the sedimentation rate and decompaction of layers in basin modelling. Shapes of the NPP can decode compaction mechanisms. Sediments buried under normal gravity force, with mechanical compaction process or exponential decrease of porosity with depth, show concave-downwards profiles (Korvin 1984; Burrus 1998) whereas concave upwards NPP are the result of pressure solution models. Here as the temperature increases, the rate of porosity loss also rises (Angevine and Turcotte 1983; Burrus 1998). The validity of these curves have been debated since some assumed an exponential porosity versus depth curve for their models based on Athy's

(1930a) work (Dutta 1987; Ungerer et al. 1990; Forbes et al. 1992) despite a linear trend reported by Hedberg (1936). We now review the Normal Porosity Profile for shales, shaly sandstones and clean sandstones. In subsequent chapters anomalous porosity profiles will be discussed.

1.3.2 Shales

Athy's (1930a) widely used exponential relation showed decrease of porosity from 0.5 to 0.05 pu, from surface up to 2.3 km depth. The normally compacted shales show concave downward patterns (e.g., Weller 1959; Perrier and Quiblier 1974; Rieke and Chilingarian 1974; Magara 1980). Other curves of shales with surface porosity ranging 0.70–0.80 pu indicates different compaction histories. As per Hedberg (1936), the normal porosity profile of shales can be divided into three parts. In the first part in the upper few hundred meters the porosity rapidly decreases to 0.35 pu. This is followed by a quasi-linear portion where porosity decreases to 0.10 pu at 2.5–3.5 km. Finally the porosity drops abruptly.

Issler (1992) in Beaufort-Mackenzie Delta (U.S.A.) showed a linear decrease of porosity from 0.35 pu at 500 m to 0.5 pu at 3700 m. The normal porosity profiles of the Nagaoka basin, Japan (Magara 1968) and the Mahakam shales (Indonesia) were also compared and all the profiles show that between 0.5 and 1 km, mechanical compaction dominates (Heling 1970) (Fig. 1.5).

The data of shales of the Amoco Lena Buerger well, Frio County, Texas shows linear increase of bulk density and decrease of porosity up to 3660 m, and constant density and porosity profile up to 5000 m (Stage 2) as reported by Powley (1993). These two stages resemble the Hedberg's compaction model explaining mechanical deformation and recrystallization. Direct porosity and density measurements also reflect these two stages. The study in the U.S Gulf Coast highlighted the normal porosity profiles for few wells in Louisiana and Texas. The porosity profiles are represented by two linear

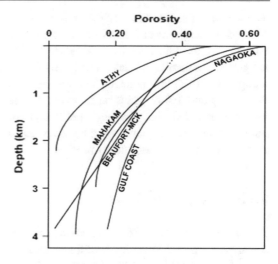

Fig. 1.5 Comparison of normal depth profile curve. Athy's original curve, shale compaction profile curve from Nagaoka basin, Gulf coast, and Beaufort Mackenzie delta. Porosity data derived from density log (Burrus 1998)

curves depicting different stages of compaction in four wells. The change in depth of occurrence from Stage 1 to Stage 2 varies from well to well. Porosity decreased steadily up to 915 m in the East Cameron well. No systematic porosity pattern exists for the next 915 m. The shale porosity profiles in other two wells, St. Landry and St. Mary, also do not show systematic decrease within 3660–4420 m depth range. The well in the Lavaca County shows linear decrease in porosity up to 2290 m and there is no compaction beyond 2290 m depth of burial indicated by nearly constant porosity. The porosity in stage 2 varies from 3 to 18% with an average of 10%. Hinch (1980) plotted the temperatures from the zone of the Stage 2 compaction in 65 wells in the Gulf coast and found that the Cretaceous sediments attained different temperatures than the Pleistocene rocks. He concluded that the temporal change in temperature processes might have played a role in initiating the Stage 2 compaction. Atwater and Miller (1965) found (i) linear decrease of porosity in sandstones, and (ii) exponential decrease in shale porosity with depth. The U.S. Gulf coast wells on the other hand show linear compaction profile of shales.

To resolve this, Bradley (1975) examined the X-ray diffractograms and undertook an elemental analysis of the shale samples. Bradley (1975) suggested dominance of clay sized quartz (67%) along with 20% clay minerals, 7% feldspar, 4% carbonates and 1% organics on an average. The author concluded that the high percentage of micron or clay sized quartz could be the possible reason for linear porosity-depth profiles.

1.3.3 Shaly Sandstones

Sandstones with 10–20% shale show concave downward normal porosity profiles similar to that of shales (Burrus 1998). Mechanical processes being the dominant compaction processes in shaly sandstones is confirmed by experiments (Rittenhouse 1971; Pittman and Larese 1991). Sediments of the NE Pacific arc related basin show similar results (Galloway 1974; Nagtegaal 1978; Burns and Ethridge 1979).

1.3.4 Sandstones

Sandstone compaction is directly proportional to the mechanical strength of the grains (Chuhan et al. 2002; Bjørlykke et al. 2009). Compaction in fine grained sand is faster than in coarse grained sand (Chuhan et al. 2002) (Fig. 1.6).

Quartz cementation is one of the important processes that enhances rock strength at >80–100 °C. This over-consolidates both sand and clay particles (Bjørlykke et al. 2004) (Fig. 1.7). Coarse-grained sandstone fracture easily by vertical/overburden stress. Chemical compaction dominates in later stages, which is directly related to temperature. Cathodeluminescence images of rock samples of the Tilje Formation (Haltenbaken, offshore Norway) shows quartz cements in fractures and at places chlorite formation (Chuhan et al. 2002).

Mechanical compaction seems to dominate overpressure solution up to first one km depth for sandstones (e.g., Tada and Siever 1989). Grains

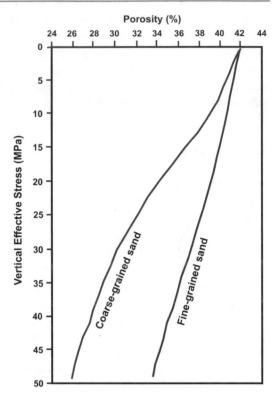

Fig. 1.6 Mechanical compaction in sands. Fine grained sands compact easily than core grained sand (Chuhan et al. 2002)

rearrange with burial leading to reduction in porosity ~ 0.26 pu. Houseknecht (1987) stated that the surface porosity of sandstones to be ~ 0.40 pu, and pressure solution begins at ~ 1.5 km depth. Pressure solution can reduce the porosity to 0.03 pu within 3–5 km. Various authors (Schmoker and Gautier 1988) also attempted the relationship between the timing of burial and temperature in terms of several indices, but no definitive physical laws were inferred (Deming 1990).

1.4 Carbonates

At low temperatures carbonates are more affected by diagenesis than clastics. Cementation and pressure solution are the two chemical processes involved in carbonate diagenesis (Bjørlykke et al.

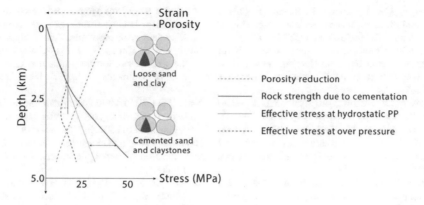

Fig. 1.7 Changes in sand compaction with burial (Bjørlykke et al. 2004)

2009). Incidentally, dissolution of aragonite components is the third important process. Dissolution, aragonite and magnesium calcite to low-Mg calcite conversion, known as mineralogical stabilization, are common in carbonate diagenesis. Pressure solution is another important process in carbonate diagenesis. Croizé et al. (2010, and references therein) compared the natural compaction trend of carbonates to that of the experimental findings. The results show that the porosity loss in carbonates is solely not because of mechanical compaction, but other processes such as chemical digenesis could be important. See Nader (2017) for case histories on carbonate diagenesis.

References

Allen PA, Allen JR (2014) Basin analysis: principles and applications to petroleum play assessment, 3rd edn. Wiley-Blackwell, Hoboken, New Jersey. ISBN 978-0-470-67377-5

Angevine CL, Turcotte DL (1983) Porosity reduction by pressure solution: a theoretical model for quartz arenites. Geol Soc Am Bull 94:1129–1134

Athy LF (1930a) Density, porosity, and compaction of sedimentary rocks. AAPG Bull 14(1):1–24

Athy LF (1930b) Compaction and oil migration. AAPG Bull 14:25–35

Atwater GI, Miller EE (1965) The effect of decrease of porosity with depth on future development of oil and gas reserves in south Louisiana. AAPG Bull 49:334

Bjørlykke K, Chuhan F, Kjeldstad A, Gundersen E, Lauvrak O, Høeg K (2004) Modelling of sediment compaction and fluid flow during burial in sedimentary basins. Mar Pet Geol 14:267–276

Bjørlykke K, Jahren J, Mondol NH, Marcussen O, Croize D, Christer P, Thyberg B (2009) Sediment compaction and rock properties. In: AAPG international conference and exhibition

Bradley JS (1975) Abnormal formation pressure. AAPG Bull 59:957–973

Burns LK, Ethridge FG (1979) Petrology and diagenetic effects of lithic sandstones-Paleocene and Eocene Umpqua Formation, southwest Oregon. In: Scholle PA, Schluger PR (eds) Aspects of diagenesis. Society of Economic Paleontologists and Mineralogists Special Publication, No 26, pp 307–317

Burrus J (1998) Overpressure models for clastic rocks, their relation to hydrocarbon expulsion: a critical reevaluation. In: Law BE, Ulmishek GF, Slavin VI (eds) Abnormal pressures in hydrocarbon environments: AAPG Memoir 70, pp 35–63

Burst JF (1969) Diagenesis of Gulf Coast clayey sediments and its possible relation to petroleum migration. AAPG Bull 53:73–93

Bushby C, Azor A (2012) Tectonics of sedimentary basins: recent advances. Wiley-Blackwell, pp 1–647

Chapman RE (1994) Abnormal pore pressures: essential theory, possible causes, and sliding. In: Fertl WH, Chapman RE, Hotz RF (eds) Studies in abnormal pressures. Developments in petroleum science, vol 38. Elsevier, pp 51–91

Chuhan FA, Kjeldstad A, Bjorlykke K, Hoeg K (2002) Porosity loss in sand by grain crushing; experimental evidence and relevance to reservoir quality. Mar Pet Geol 19(1):39–53

Croizé D, Bjorlykke K, Dysthe DK, Renard F, Jahren J (2008) Deformation of carbonates, experimental mechanical and chemical compaction. Geophysical Research Abstracts 10

Croizé D, Bjørlykke K, Jahren J, Renard F (2010) Experimental mechanical and chemical compaction of carbonate sand. JGR Solid Earth 115:B11204

Deming D (1990) Comment on "Compaction of basin sediments: modeling based on time-temperature history" by J.W. Schmoker and D.L. Gautier. J Geophys Res 95:5153–5154

Dutta NC (1987) Geopressure, geophysics reprint series no. 7, Society of Exploration Geophysicists

Forbes PL, Ungerer P, Mudford BS (1992) A two dimensional model of overpressure development and gas accumulation in Venture Field, eastern Canada. AAPG Bull 76(3):318–338

Galloway WE (1974) Deposition and diagenetic alteration of sandstone in northeast Pacific arc-related basin: implications for graywacke genesis. Geol Soc Am Bull 85:379–390

Hamilton EL (1959) Thickness and consolidation of deep-sea sediments. Geol Soc Am Bull 70:1399–1424

Hedberg HD (1936) Gravitational compaction of clays and shales. Am J Sci 31:241–287

Heling D (1970) Micro-fabrics of shales and their rearrangement by compaction. Sedimentology 15:247–260

Hinch HH (1980) The nature of shales and the dynamics of hydrocarbon expulsion in the Gulf Coast Tertiary section. In: Roberts WH III, Cordell RJ (eds) Problems of petroleum migration: AAPG Studies in geology, no 10. Tulsa, The American Association of Petroleum Geologists, pp 1–18

Houseknecht D (1987) Accessing the relative importance of compaction processes and cementation to reduction of porosity in sandstones. AAPG Bull 71:633–642

Hunt JM, Whelan JK, Eglinton LB, Cathles LM III (1998) Relation of shale porosities, gas generation, and compaction to deep overpressures in the U.S. Gulf Coast. In: Law BE, Ulmishek GF, Slavin VI (eds) Abnormal pressures in hydrocarbon environments: AAPG Memoir 70, pp 87–104

Issler DR (1992) A new approach to shale compaction and stratigraphic restoration, Beaufort-Mackenzie Basin and Mackenzie Corridor, Northern Canada. Am Assoc Pet Geol Bull 76:1170–1189

Korvin G (1984) Shale compaction and statistical physics. Geophys J R Astron Soc 78:35–50

Magara K (1968) Compaction and migration of fluids in Miocene mudstone, Nagaoka Plain, Japan. AAPG Bull 52:2466–2501

Magara K (1980) Comparison of porosity depth relationships of shale and sandstone. J Pet Geol 3:175–185

Mukherjee S (2017) Airy's isostatic model: a proposal for a realistic case. Arab J Geosci 10:268

Mukherjee S (2018a) Locating center of pressure in 2D geological situations. J Indian Geophys Union 22:49–51

Mukherjee S (2018b) Locating center of gravity in geological contexts. Int J Earth Sci 107:1935–1939

Mukherjee S (2018c) Moment of inertia for rock blocks subject to bookshelf faulting with geologically plausible density distributions. J Earth Syst Sci 127:80

Mukherjee S, Kumar N (2018) A first-order model for temperature rise for uniform and differential compression of sediments in basins. Int J Earth Sci 107:2999–3004

Nader FH (2017) Multi-scale quantitative diagenesis and impacts on heterogeneity of carbonate reservoir rocks. Springer. ISBN 978-3-319-464-45-9

Nagtegaal PJC (1978) Sandstone-framework instability as a function of burial diagenesis. J Geol Soc London 135(1):101–105

Overton HL, Zanier AM (1970) Hydratable shales and the salinity high enigma. Fall Meeting of the Society of Petroleum Engineers of AIME. Society of Petroleum Engineers, Houston, TX, Pap. 2989, 9 pp

Perrier R, Quiblier J (1974) Thickness changes in sedimentary layers during compaction history; methods for quantitative evaluation. AAPG Bull 58:507–520

Pettijohn FJ (1984) Sedimentary rocks, 3rd edn. CBS Publishers and Distributors, p 58

Pittman ED, Larese RE (1991) Compaction of lithic sands: experimental results and applications. AAPG Bull 75:1279–1299

Powers MC (1967) Fluid-release mechanism in compacting marine mud-rocks and their importance in oil exploration. AAPG Bull 51:1240–1245

Powley DE (1993) Shale compaction and its relationship to fluid seals. Section III, Quarterly report, Jan 1993–Apr 1993, Oklahoma State University to the Gas Research Institute

Rieke HH, Chilingarian GV (1974) Compaction of Argillaceous Sediments. Elsevier, New York, p 424

Rittenhouse G (1971) Mechanical compaction of sands containing different percentages of ductile grains: a theoretical approach. AAPG Bull 55:92–96

Schmoker JW, Gautier DL (1988) Sandstone porosity as a function of thermal maturity. Geology 16:1007–1010

Tada R, Siever R (1989) Pressure solution during diagenesis. Annu Rev Earth Planet Sci 17:89–118

Teodorovich GI, Chernov AA (1968) Character of changes with depth in productive deposits of Apsheron oil-gas-bearing region. Soviet Geol 4:83–93

Ungerer P, Burrus J, Doligez B, Chenet P-Y, Bessis F (1990) Basin evaluation by integrated two dimensional modeling of heat transfer, fluid flow, hydrocarbon generation and migration. AAPG Bull 74:309–335

Warner DL (1964) An analysis of influence of physical-chemical factors upon the consolidation of fine grained elastic sediments. Thesis, University of California, 136 pp

Weller EA (1959) Compaction of sediments. AAPG Bull 43:273–310

Porosity in Carbonates

2

Abstract

Porosity in carbonate rocks, most commonly limestones and dolostones, is of great importance to study since around half of world's hydrocarbon reserves are made up of dolomite and limestone, which formed mostly in a shallow marine environment and usually close to where such sediments originate from the source rocks. Carbonates possess both primary and secondary porosities, which reduces with progressive burial leading to increasing rigidity of the rock. Several classifications of carbonate rocks are available. These are based on texture, depositional environments (the three kinds of carbonate factories), energy of the depositional environment, mud to grain ratio (volume-wise), grain to micrite ratio, porosity-permeability parameters, depositional-, diagenetic- and biological issues etc. Out of them, those by Folk and Dunham have been entered most of the text books on sedimentology. Carbonates more commonly display dissolution, cementation, recrystallization and grain replacement than the siliciclastic deposits. The porosity-permeability relation in carbonates may or may not be linear. Several schemes of classification of porosity of carbonates are available. Archi's scheme (based on qualitative evaluation of texture and porosity), the Choquette-Pray scheme (utilizes depositional and diagenetic changes in the rock), the Lucia scheme (works on inter-relationship between porosity, permeability and the particle size) etc.

2.1 Introduction

Nearly 50% of world's oil and gas reserves are in dolomite and limestone. Dolomites are often more porous than limestones. Carbonates are characterised by multi-porosity characteristics unlike sandstones. Carbonate rocks exhibit various types of pores starting from primary porosity formed at the time of deposition to secondary porosity resulting from diagenesis. Cementation leads to porosity destruction and with time and burial the general trend in pore systems is towards destruction, but there are certain processes that preserve porosity in the overpressured formation. With progressive burial, porosity falls along with increase in framework grain rigidity.

This chapter discusses the followings: (i) the basic characteristics of carbonate sediments, (ii) their genesis, (iii) changes in their fabric, (iv) depositional processes, (v) the ability of certain organisms to build structures and their relevance. This chapter discusses how porosity evolves and diagenetic changes takes place in carbonates.

T. Dasgupta and S. Mukherjee, *Sediment Compaction and Applications in Petroleum Geoscience*, Advances in Oil and Gas Exploration & Production, https://doi.org/10.1007/978-3-030-13442-6_2

2.2 Origin of Carbonate Rocks

Studies by Milliman (1974), Wilson (1975), Tucker and Wright (1990) and Moore (2001) indicated that the modern day carbonates are mainly biotic and form mostly in (shallow) marine environment. The formation of carbonates depends on the parameters favourable for carbonate deposition such as suitable temperature, salinity, and presence of hard substrate and absence of siliciclastics (Lees 1975; Moore 2001). Figure 2.1 shows the latitude-wise distribution and abundance of organisms. The growth of most of the corals (i.e., besides the cold water corals) mainly depends on the presence of light, so there is prolific growth of carbonates in the upper part of the marine environment up to ~10 m depth (Moore 2001; Fig. 2.2).

The study on the origin of carbonates with coarser grains can be commented by observing shell fragments, entire foraminifera etc. Genesis of carbonate mud involves several processes including particles derived from erosion of the shells of typically tropical climate (Moore 2001). From the dead green algae, deposited aragonite needles finally produce carbonate mud (Moore 2001). The important aspect of carbonate

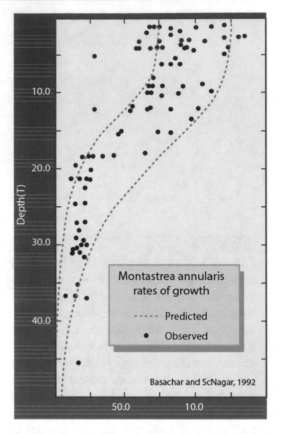

Fig. 2.2 The dashed line represents the predicted growth and the open circles represent the actual growth of the corals (Moore 2001)

Fig. 2.1 Latitude-wise distribution of organisms (Moore 2001)

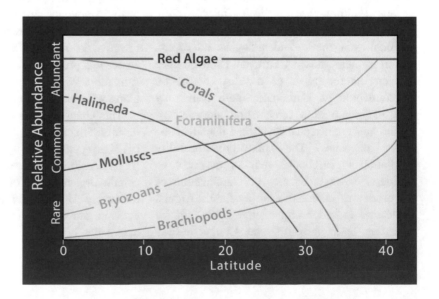

sedimentation is that unlike siliciclastics, they deposit close to where they originate (Moore 2001).

2.3   Carbonate Classification

The carbonate classifications in the early century was not a priority. With time, however, main issues were addressed: carbonates can deposit in (i) quiet water; and as (ii) current-washed deposits. In this classification calc-arenites and fine-grained limestones were clubbed in the quiet water group and the coarse grained clastics into the current-washed categories. The Upper Jurassic Arabian limestones were classified into different types based on original texture and the abundances of mud/grain ratio. The different types are: (i) aphanitic limestones (<10% particles); (ii) calc-arenitic limestones (mud with >10% particles); (iii) calc-arenites (sand with <10% mud matrix); (iv) coarse clastic carbonates (gravel with <10% mud matrix); and (v) residual organic limestones (in situ reef rocks).

Plumley et al. (1962) classified limestones based on the type of energy of the depositional environment. These are: (i) quiet water; (ii) intermittently agitated water; (iii) slightly agitated water; (vi) moderately agitated water; and (v) strongly agitated water.

Leighton and Pendexter (1962) classified carbonates based on grain to micrite ratio (Mazzullo and Chilingarian 1992). Micrite being very fine-grained, it was characterised by some workers as the place where the sedimentary particles are embedded (Plumley et al. 1962). Dunham (1962) classified micrites as particles with <62 μm size. The organic structures and recrystallization fabrics were identified by Leighton and Pendexter (1962) and his classification is as follows: (i) micritic limestones; (ii) detrital limestones with embedded conglomerates of older limestone units; (iii) skeletal limestones; (iv) pellet limestones; (v) lump limestones; (vi) limestones with coated grains including oolite, pisolites etc. and (vii) mainly reefal limestones.

Leighton and Pendexter (1962) classification scheme was modified by Bissell and Chilingar (1967) according to micrite to grain ratio. Another scheme by Thomas (1962) where the grain particles and the cement and porosity-permeability parameters were also considered in the classification of Paleozoic limestones is as follows: (i) skeletal part, (ii) non-skeletal part, (iii) organic matter, and (iv) breccia.

Later, Folk's (1959, 1962) and Dunham's (1962) classification became very popular (Ahr 2008). Riding (2002) classified reefs and discussed their geneses.

Folk's and Dunham's classifications have some similarities. They depend mainly on the (volume) ratio between the mud and grains, and the packing arrangement of the grains. The pattern of textural maturity plays a major role in sandstone description and similar concepts have been applied in carbonate classification both by Folk and Dunham (Ahr 2008). Rocks with >90% lime mud were designated as mudstone by Dunham and as micrite by Folk. The rocks where grainstones are dominant are referred as sparite by Folk. Dunham referred these rocks as grainstones. Depending on the proportion of the constituent grains and their packing, the rocks are named differently. Commonly Dunham's classification is used in industry for carbonate classification. This is because the reservoir properties can be framed easily from the rock description. The environment of deposition can be easily interpreted from the rock types, as the mudstones form where the winnowing is insignificant. Rocks with high grain percentage are deposited in high energy environment. Grainstones and packstones have highest intergranular porosity and these rock types are also prone to diagenetic alteration leading to early cementation and decrease in pore throat size.

2.4   Carbonate Factories

The carbonate factories are clubbed into three broad categories: T (for tropical, top water), M (mud mound, micrite) and C (cool water and mainly biogenic precipitation) (Fig. 2.3, Schlager 2005). These carbonate factories differ

Fig. 2.3 Three carbonate
factories with different
mineralogical contents
(Schlager 2005)

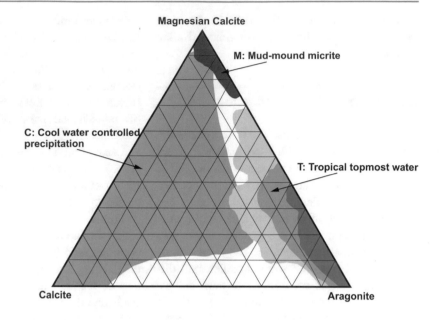

Fig. 2.4 Different modes of
precipitation of carbonate
factories (Schlager 2005)

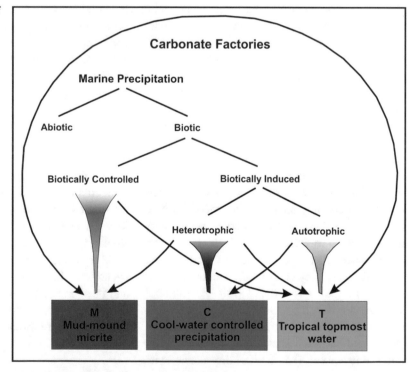

in many aspects such as mineralogical content
(Fig. 2.3), precipitation mode (Fig. 2.4), different
depth ranges, and production potentials.

The T carbonates are generally formed within
the 30°N to 30°S latitude in ~20 °C warm

waters. Reef building corals and some molluscs
are the common sediment building organisms
that are mainly photosynthetic symbionts (Moore
and Wade 2013). Abiotic components of carbonates such as ooids and whitings are also

common and are composed of aragonite and magnesian calcite (Schlager 2005).

The C carbonates/cool water carbonates form nd with prolific supply of nutrients from cold waters (Schlager 2005). The sediment producers are mainly heterotrophs and cement precipitators. Mineralogy of such carbonates is dominantly calcite since in cold waters aragonite and magnesian calcite may dissolve (Fig. 2.3). The production is insensitive to light and thus may occur in deep water (Fig. 2.5).

The M carbonates/mud mound consists of micritic calcite muds. Mud mounds were prolific

during the Paleozoic. Abiotic cement comprises of mud mounds forming stromatactis fabric (Schlager 2005). It forms in the low-light intensity zone and the mineralogy is commonly calcite (Fig. 2.3).

Cations (Ca^{2+}, Mg^{2+}, Fe^{2+}, Mn^{2+}, Zn^{2+}, Ba^{2+} and Sr^{2+}) and anionic complexes form carbonates. Depending on the crystal lattice structure carbonates can be grouped into different families. Hexagonal, orthorhombic and monoclinic are the common crystallographic systems that represent these families. The common minerals are calcite (hexagonal system) and aragonite (orthorhombic

Fig. 2.5 Depth of occurrence of carbonate factories along with production rate (Schlager 2005)

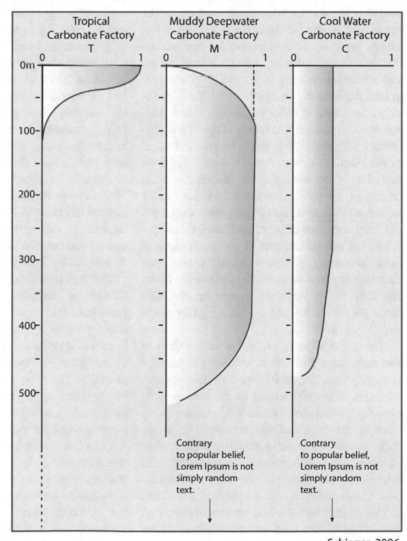

Schiager, 2006

system). In dolomites there is loss in rotational symmetry as the Mg ions, smaller than the Ca ions, enter the structure and alter the lattice. Limestones and dolostones commonly constitute carbonates (Reeder 1983).

The texture of siliciclastic rocks strongly depend on the type of parent rock, weathering type and the transportation duration. Such rocks are composed of quartz, feldspar, rock fragments and matrix. Ham and Pray (1962) first identified the distinct attributes of carbonate rocks. Biological, chemical and detrital processes may involve in carbonate formation. Their mineralogy is independent on the composition of the weathered and parent rocks, and their texture is also depends on the flow stream patterns of the rivers unlike the siliciclastics. Carbonates are mainly made up of skeletal remains and biological constituents such as lime mud, fecal pellets and microbial cements (Folk 1980). Siliciclastic grains do exist in carbonates and these could either be clasts of older rocks or lithified fragments or reworked sediments. Their main difference with the siliciclastics is that the carbonate development can depend on biological activities and many times the carbonate stratifications are destroyed by the burrowing activities. Finally unlike siliciclastics, carbonates alter diagenetically by rapid dissolution, cementation, recrystallization and replacement of grains. Because of these secondary processes fractures are more common in carbonates unlike siliciclastics (Ham and Pray 1962). Major differences between siliciclastics and carbonates are listed by Choquette and Pray (1970).

The petrophysical properties of sandstones and carbonates differ. In sandstones the porosity is mainly classified as inter-particle and the permeability is closely related to the inter-particle porosity. The petrophysical lab measurements made on few inches of core represents the entire rock volume. Carbonates exhibit a wide variety of pore types with varying sizes, shapes and origins. In this case, porosity does not always bear a linear relation with the permeability. Thus in case of carbonates, lab measurements made on few inches of core does not always represent the entire facies. To improve the situation, one foot

(0.305 m) long core would be required for reliable correlation between the core and the facies (Ahr 2008).

2.5 Porosity and Its Classification

The porosities of Holocene carbonate sediments are very high: ~ 40–75% and these higher values are common in micritic limestones (Fig. 2.6; Moore 1989). High porosities are associated in the deep water facies that are mainly oozes and these have both inter-and intra-particle porosities (Schlanger and Douglas 1974). Reefs represent that section of the carbonates where the framework pores represent a large portion of porosity and permeability in reefal rocks.

In recent unconsolidated carbonates, porosity arises mainly by: (i) boring processes in reefs due to advent of algae, bacteria or fungi. Bivalves and sponges show boring features in hard bases; (ii) animals and plants show bioturbation; (iii) fenestral structures formed by micro-organisms; and (iv) dissolution feature such as evaporate dissolution.

Porosity in modern carbonates ranges 40–70% (Choquette and Pray 1970), whereas in lithified old samples it is merely 5–15%. Porosity reduces in carbonates mainly by compaction and/or cementation (also see Mukherjee and Kumar 2018). Porosity reduces in ancient rocks mainly by cementation and pressure dissolution. Commonly carbonates do not have regular pores/pore throats and in order to identify the rock properties, the porosity classification becomes important (Ahr 2008). The common classification is inter-particle and secondary porosities like vugs and fractures. Another way of classification is grouping into shape, size of pore, rock properties i.e. mainly petrophysical properties and the mode of origin. Rock typing is a common methodology to characterise the reservoir where the porosity, permeability and pore throat sizes are linked to classify according to hydraulic units (Ahr 2008). The evolution of the carbonate reservoir pore systems are very well explained in the schemes by Archie (1952), Choquette and Pray (1970) and Lucia (1983).

Fig. 2.6 Porosity
classification incorporating
the details about the
depositional as well as the
diagenetic changes and are as
categorised as fabric selective,
not fabric selective and fabric
selective or not category
(Scholle 1978)

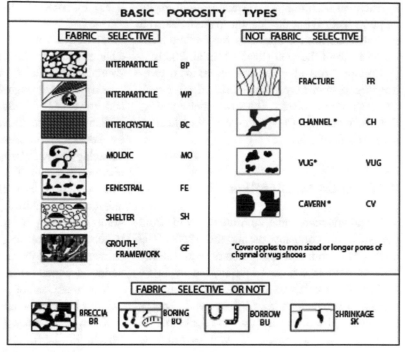

2.6   Porosity Classification

2.6.1   Archie's Scheme

Carbonates were first classified by Archie (1952). Archie's classification was primarily based on textural description and the type of visible porosity. Textural categories are from type I to III and are based on visible porosities of four classes from A to D. At a magnification of 10×, class A has no visible porosity, any porosity between 1 and 10 μm is referred to as class B. Class C has a visible porosity of more than 10 μm.

Type I carbonates are referred as crystalline, dense, hard and under a microscope it shows no visible porosity. The Solenhofen Limestone (Germany) is the example of Type I carbonates (Ahr 2008). Type II porosity are mostly chalky and wackestone type and the pore size does not exceed 50 μm. The granular carbonates are mostly referred as type III carbonates and they come under the grainstones and packstones categories (Ahr 2008). Archie's classification follows an integrated approach where the capillary pressures, electrical properties and the saturation characteristics of the rock type were integrated. The composition of the rock, its mineralogical content as well as the provenance was not considered into Archie's classification (Ahr 2008). It was difficult to use Archie's porosity classification to address and relate porosities along with their genesis.

2.6.2   Choquette-Pray Classification

In the Choquette and Pray classification (1970), 15 types of pores were categorised into three subgroups: (i) fabric selective, (ii) non-fabric selective, and (iii) may or may not be fabric selective. In this type of classification, all the information including the depositional and diagenetic changes is incorporated. The main examples of fabric selective pores are oomoldic porosities and intercrystalline porosity as encountered in dolomites. Grain moldic pores are mainly intergranular. Non-fabric selective pores fall mainly into category of fractures or dissolution cavities that cut across the fabrics.

Mechanical stratigraphy and fracture stratigraphy helps in identifying the fracture patterns as well as understanding the attributes for fracture studies away from the well (Laubach et al. 2009).

Certain types of porosities represent a category that may or may not be fabric selective such as desiccation cracks, burrows and boring. Details about the Choquette-Pray classification are given in Fig. 2.6.

2.6.3   Lucia Classification

This classification came up after the work from the Shell oil company during the 1960s. The inter-relationship amongst porosity, permeability and the particle size was investigated by Lucia (1995). The Lucia classification allowed to distinguish between inter-particle and vuggy porosities. Also different vugs were distinguished based on their separable as well as touching characteristics. Craze (1950) and Bagrintseva (1977) studied the relationship between porosity, permeability, some aspects of capillary pressure and particle size. They studied the main rock types using this classification. In Lucia classification the main work was to delineate petrophysically inter-particle pores and vugs. The porosity is classified in terms of particle size as fine- (<20 μm), medium- (20–100 μm) and large pores (>100 μm). The classification mainly characterises the pores in limestones, dolostones and mudstones. The geological origin about the pore spaces were not explained by Lucia's classification whereas the highlight of this classification is the inter-relationship between rocks and their petrophysical properties. Further, Lønøy (2006) subdivided pores into inter-particle and inter-granular types and this scheme better correlates porosity and permeability.

At places where the sediments are deposited and the pores formed contemporaneously, a genetic correlation between the primary pores and the rock facies can be made (Ahr 2008). The genetic classification between secondary pores such as vugs and fractures and the rock facies seems not possible.

2.7   Permeability Classification

Much of the work was done by Henri Darcy and Charles Ritter on the flow of water through sands around 1856. The flow rate was determined by passing water through the cylinder made of sands and gravels (Todd and Mays 2012).

Same as porosity, permeability can develop by several processes. Processes involved in changing the rock properties may act more than once in the geological history. Sorting and grain size variation are the important properties affecting permeability (Ahr 2008). As the smaller grains enter the pore spaces, the pore throat clogs thereby decreasing permeability. The deliverability/economics of a carbonate reservoir mainly depends on the permeability.

The main factor that enhances the permeability of the carbonate reservoir is the presence of fracture porosity (Watts 1983; Tucker and Wright 1990). In general, lithologies like chalk that are fine-grained have low permeability but certain processes such as re-sedimentation that control mineral concentration enhances the permeability of the matrix. Radial and concentric fractures resulting from doming/diapirism also increase the flow network as seen at the doming of Zechstein evaporites (Watts 1983).

2.8   Diagenetic Processes and Porosity Development

Diagenetic processes involve dissolution that increases porosity, recrystallization, and replacements of minerals, mineralogical transformation, evaporitization and cementation (Mazzullo and Chilingarian 1992). Diagenesis is broadly defined as the chemical and textural changes, which occur in rocks during post-depositional procedure during the contact of active fluids in the whole process (Mazzullo and Chilingarian 1992). Compositionally these fluids are different: marine, brackish, normally saline, or hypersaline.

The diagenetic phases viz. eogenetic, mesogenetic and telogenetic stages correlate to various physical or chemical or both processes (Choquette and Pray 1970). In eogenetic stage, diagenesis of sub-aerially exposed marine sediments involving meteoric water takes place (Mazzullo and Chilingarian 1992). As explained by Harris et al. (1985), the mesogenetic diagenetic stage corresponds mainly to the burial diagenesis zone where the porosity changes and the diagenesis results in the change in bulk volume. Telogenetic diagenesis occurs by weathering of old carbonate rocks after uplift. This causes porosity formation in subaerial unconformity zones (Mazzullo and Chilingarian 1992).

References

Ahr WM (2008) Geology of carbonate rocks. Wiley Publication, New York

Archie GE (1952) Classification of carbonate reservoir rocks and petrophysical considerations. AAPG Bull 36:278–298

Bagrintseva KI (1977) Carbonate rocks, oil and gas reservoirs. Izdated'stvoNedra, Moscow, 231 pp

Bissell HJ, Chilingar GV (1967) Classification of sedimentary carbonate rocks. In: Carbonate rocks-origin, occurrence and classification. Elsevier Publishing Company, Amsterdam-London-New York, pp 87–168

Choquette PW, Pray LC (1970) Geological nomenclature and classification of porosity in sedimentary carbonates. AAPG Bull 54:207–250

Craze RC (1950) Performance of limestone reservoirs. J Petrol Technol 189:287–294

Dunham RJ (1962) Classification of carbonate rocks according to depositional texture. In: Classification of carbonate rocks. In: Ham WE (ed) Classification of Carbonate Rocks–A Symposium. AAPG Memoir, pp 108–121

Folk RL (1959) Practical petrographic classification of limestones. AAPG Bull 43:1–38

Folk RL (1962) Spectral subdivision of limestone types. In: Ham WE (ed) Classification of carbonate rocks. AAPG Memoir No 1, Tulsa, OK, pp 62–84

Folk RL (1980) Petrology of sedimentary rocks. Hemphill Publishing Company

Ham WE, Pray LC (1962) Modern concepts and classifications of carbonate rocks. In: Classification of carbonate rocks. AAPG Memoir No 1, pp 2–19

Harris PM, Christopher G, Kendall C, Lerche I (1985) Carbonate cementation—a brief review, Society for Sedimentary Geology, Vol 36. https://doi.org/10.2110/pec.85.36.0079

Lønøy A (2006) Making sense of carbonate pore systems. AAPG Bull 90:1381–1405

Lucia FJ (1983) Petrophysical parameters estimated from visual descriptions of carbonate rocks: a field classification of carbonate pore space. J Pet Technol 35:629–637

Mazzullo SJ, Chilingarian GV (1992) Diagenesis and origin of porosity. In: Chilingarian GV, Mazzullo SJ, Rieke HH (eds) Carbonate reservoir characterization: a geologic-engineering analysis, Part I: Elsevier Publ. Co., Amsterdam, Developments in Petroleum Science 30, pp. 199–270

Milliman JD (1974) Recent sedimentary carbonates. Springer-Verlag, Berlin, p 375

Moore CH (1989) Carbonate diagenesis and porosity. Elsevier, Amsterdam

Moore CH (2001) Carbonate reservoirs: porosity evolution and diagenesis in a sequence stratigraphic framework. Elsevier, Amsterdam, p 444

Moore C, Wade WJ (2013) Carbonate reservoirs: porosity and diagenesis in a sequence stratigraphic framework, 2nd edn, vol 67

Mukherjee S, Kumar N (2018) A first-order model for temperature rise for uniform and differential compression of sediments in basins. Int J Earth Sci 107:2999–3004

Laubach SE, Eichhubl P, Olson JE (2009) Fracture diagenesis and producibility in tight gas sandstones. In: Carr T, D'Agostino T, Ambrose W, Pashin J, Rosen NC (eds) Unconventional energy resources: making the unconventional conventional. 29th Annual GCSSEPM Foundation Bob F. Perkins research conference, pp 438–499

Lees A (1975) Possible influence of salinity and temperature on modern shelf carbonate sedimentation. Mar Geol 19:159–198

Leighton MW, Pendexter C (1962) Carbonate rock types. In: Ham WE (ed) Classification of carbonate rocks, American Association of Petroleum Geologists, Mem. 1, pp 33–61

Lucia FJ (1995) Rock-fabric/petrophysical classification of carbonate pore space for reservoir characterization. AAPG Bull 79(9):1275–1300

Plumley WJ, Risley GA, Graves Jr, Kaley ME (1962) Energy index for limestone interpretation and classification. American Association of Petroleum Geologists. https://doi.org/10.1306/M1357

Reeder RJ (1983) Carbonates: mineralogy and chemistry. Reviews in Mineralogy, vol 11. Mineralogical Society of America, 394 pp

Riding R (2002) Structure and composition of organic reefs and carbonate mud mounds; concepts and categories. Earth Sci Rev 58:163–231

Schlager W (2005) Carbonate sedimentology and sequence stratigraphy. SEPM Concepts in Sedimentology and Paleontology. SEPM, Tulsa. ISBN 1-56576-116-2

Schlanger SO, Douglas RG (1974) The pelagic oozechalk-limestone transition and its implications for marine stratigraphy. Pelagic Sed 1:117–148

Scholle PA (1978) A Color Illustrated Guide to Carbonate Rock Constituents, Textures, Cements, and Porosities. American Association of Petroleum Geologists, Memoir 27. xiii + 241 pp., numerous colour plates. Tulsa, Oklahoma. IJF 115(6):473. https://doi.org/10.1017/S0016756800041881

Thomas GE (1962) Grouping of carbonate rocks into textural and porosity units for mapping purposes. In: Ham WE (ed) Classification of carbonate rocks—a symposium. Am. Assoc. Petrol. Geol. vol 1, p 193

Todd DK, Mays LW (2012) Groundwater hydrology, 3rd edn. Wiley-India Edition, pp 86–145

Tucker M, Wright VP (1990) Carbonate sedimentology. Blackwell Scientific, Oxford, p 482

Watts NL (1983) Microfractures in chalks of Albuskjell Field, Norwegian Sector, North Sea: possible origin and distribution. AAPG Bull 67:201–234

Wilson JL (1975) Carbonate facies in geologic history. Springer-Verlag, New York, p 471

Pore Pressure Determination Methods

Abstract

Overpressure situation can be created in both clastic and non-clastic reservoirs when at some depth the formation pressure exceeds what is expected for a hydrostatic (normal/lithostatic) pressure scenario. Likewise an underpressure situation has also been reported from reservoirs after sufficient hydrocarbons have been extracted. Over- and underpressure can develop by both tectonic (e.g., horizontal or vertical stress) and atectonic processes (e.g., mineral phase change, kerogen maturation). Presence or withdrawal of water (saline and freshwater) and hydrocarbon can produce over- and underpressure. Fracture pressure and its gradient are important in planning well-drilling programmes. Pore pressure estimation has become an active field of research in the present day oil industry and several methods exist for such estimation.

3.1 Introduction

This chapter mainly deals with the basic aspects required for pore pressure understanding/calculations starting from normal pressure to abnormal pressure scenarios. The estimation of pore pressure is important for well planning. In wild cat exploratory wells pore pressures are estimated using the seismic velocities whereas in known areas, offset well data including information of logs, cuttings, seismic velocities help in building pore pressure curves to avoid events such as blow-outs while drilling.

In porous formations, the fluid pressures in the pore spaces define the pore pressure. There is a huge variation in pore pressure, right from hydrostatic pressure and beyond, and also in overpressure scenarios. The different causal mechanisms of pore pressure generation along with the different methodologies/techniques are highlighted in this chapter.

3.2 Normal Pressure

Hydrostatic/lithostatic/normal pressure (Zhang 2011) is the pressure exerted by the height of column of water/rockon the formation under consideration (Donaldson et al. 2002).

$$\rho_h = \gamma_w \times h \qquad (3.1)$$

ρ_h hydrostatic pressure (lb/ft^2)
γ_w specific weight of water (lb/ft^3)
h height of column of water (feet).

The hydrostatic pressure gradient is referred to as G_h

$$G_h = \gamma_w / 144 \qquad (3.2)$$

Abnormal pore pressure is a pressure that differs from the hydrostatic pressure whereas overpressure is the situation when the formation pore pressure exceeds normal pressure (Zhang 2011). The pore pressure prediction principle works based on Terzaghi's and Biot's effective stress relations (Biot 1941; Terzaghi et al. 1996). According to this theory, the pore pressure is a function of total stress or overburden stress and effective stress. The simple expression relating overburden stress, effective vertical stress and pore pressure is

$$P = (\sigma_v - \sigma_e)/\alpha \qquad (3.3)$$

P   pore pressure
σ_v overburden stress
σ_e effective stress
α the Biot effective stress coefficient. Commonly $\alpha = 1$ is taken (Zhang 2011).

The combined pressure effect of grain to grain matrix and the fluid present in it causes the pressure to be 1 psi/ft (0.2351 kg/cm^2/m). According to Swarbrick and Osborne (1998) the lithostatic pressure gradient reaches 0.7 psi/ft at very shallow depth with 60–70% porosity.

If the overburden stress and the effective stress are known pore pressure can be calculated by Eq. 3.3. In relatively younger basins the pore pressure profile resembles that of Fig. 3.1 where up to 2000 m the pore pressure is nearly hydrostatic. As shown in Fig. 3.1, effective stress is the difference between overburden stress and pore pressure.

3.3   Overpressure

According to Dickinson (1953), overpressure corresponds to the formation pressure when it is more than hydrostatic pressure with water or brine in the formation (Fig. 3.2). In a "disequilibrium state", the fluid retention capability of the impermeable beds leads to overpressure scenarios (Swarbrick et al. 2002). These highly overpressured zones are often termed as the "abnormally high formation pressured zones" (AHFPs) (Donaldson et al. 2002). The pressure in such reservoirs

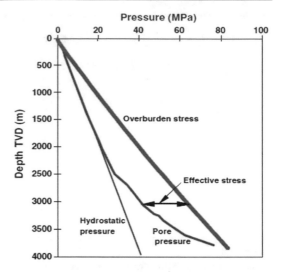

Fig. 3.1 Typical hydrostatic pressure, pore pressure, overburden stress in a borehole well (Zhang 2011)

develops when an impermeable zone exists and the AHFP zones are isolated. These barriers may be due to chemical and/or physical processes (Louden 1972). Overpressure develops by many processes such as (i) sediment compaction (Mukherjee and Kumar 2018), (ii) tectonic compression, (iii) faulting, (iv) diapirism (Mukherjee et al. 2010; Mukherjee 2011), (v) mineral phase change, (vi) kerogen maturation and hydrocarbon generation, (vii) osmosis etc. (Swarbrick and Osborne 1998; Donaldson et al. 2002). Any combination of these processes causes physico-chemical changes in pore pressure generation (Fertl 1976).

Pore pressure information is very critical for well planning and drilling. Drillers study the pore pressure gradient as it is convenient for calculating the mud weight. Three types of studies are made, i.e., pre-drill, during drilling and post well pore pressure analyses (Zhang 2011).

Worldwide overpressured zones are present in both carbonate and clastic reservoirs (Swarbrick and Osborne 1998). According to Hunt (1990), 180 basins are overpressured in America, Africa, Middle East, Far East, Europe, Australia, Asia and the age of the rocks range from Pleistocene to Cambrian. Overpressures are often associated with hydrocarbon bearing reservoirs, such as the

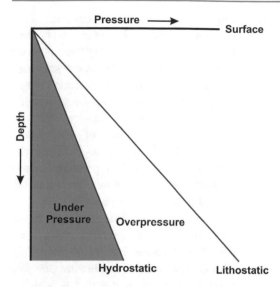

Fig. 3.2 Pressure versus depth plot showing "overpressure" zones when the pressure is more than hydrostatic. "Underpressure zones" develops when the pressure is less than the hydrostatic pressure (Swarbrick and Osborne 1998; Swarbrick et al. 2002)

deeper and highly overpressured reservoirs containing oil and gas in the Northern and Central North sea (Swarbrick and Osborne 1998). Law et al. (1998) stated that in US Gulf coast seven highly overpressured Jurassic to Recent stratigraphic zones.

3.4 Underpressure

When the pore pressure is less than the hydrostatic pressure (Fig. 3.2), underpressure scenarios develop (Swarbrick and Osborne 1998) and are also termed abnormally low formation pressures (ALFPs), (Donaldson et al. 2002). Underpressure reservoirs are mostly encountered after depletion of oil and gas after production. Naturally underpressured reservoirs also exist, particularly in Canada, U.S.A. and Alberta. In many regions, fluid withdrawal produced subsidence such as the Po delta of Italy, Bolivar coast of Lake Maracaibo (Venezuela), Galveston Bay (Texas), Taiwan, Long Beach, California, etc.

Fracture pressure is the pressure, which develops fractures in the lithological unit and

ultimately results in mud loss in the borehole (Zhang 2011). Fracture gradient is to be known during well planning and drilling, which helps to design the mud weight. Fracture gradient is obtained by dividing the true vertical depth by fracture pressure. Tensile fractures tend to form in the borehole if the mud weight exceeds the formation pressure leading ultimately to mud losses. Leak off test (LOT) helps in determining the ultimate fracture pressure and is a pre-drilling exercise.

According to Swarbrick and Osbrone (1998), overpressure formation can be explained by four aspects of rocks and fluids: (i) causal mechanisms, (ii) sealing capacity of the rock can be explained by rock permeability, (iii) fluid type, and (iv) timing of the evolution or rate of flow.

(i) **Causal mechanism**: The extent of overpressure generation mechanism depends upon causal mechanism and can be grouped into three categories:

1. Stress related mechanism, which mainly reduces the pore volume:

   (a) Vertical stress related mechanism categorized mainly as disequilibrium compaction.
   (b) Lateral stress related mechanism: mainly tectonic stresses.

2. Increase in fluid volume:

   (a) Aquathermal processes.
   (b) Oil to gas cracking.
   (c) Hydrocarbon generation resulting in an increase in fluid volume.
   (d) Mineral transformation.

3. Fluid movements, buoyancy related mechanism:

   (a) Osmosis.
   (b) Hydraulic head.
   (c) Density contrasts resulting in buoyancy.

The above mentioned causal mechanisms are explained in Table 3.1 (Swarbrick and Osborne 1998)

(ii) Sealing capacity of the rock in relation to dynamic property like permeability: Rock permeability depends on size and shape of grains and on the tortuosity of fluid flow, and on the dynamic viscosity and density (therefore also kinematic viscosity, which equals dynamic viscosity divided by density) of the fluids. Overpressure generation in non-reservoir rocks such as shale is mainly because of fluid retention in such rocks also referred as seals/cap rocks in hydrocarbon geosciences. The overpressure can be dissipated by either fracturing or connecting porous media in between. Byerlee (1993) mentioned overpressure dissipation by tectonic activities such as fault reactivation. Hydraulic fracturing is possible in rocks if the overpressure extent reaches the fracture gradient and releases the amount of stored fluid that ultimately releases the pressure (Engelder 1993).

(iii) **Fluid type**: The most common fluid occurring in nature is water, which could be either fresh or saline. Hydrocarbons are always associated with water and their flow depends on the hydrocarbon composition, temperature in the in situ conditions, hydrocarbon saturation and the relative permeabilities of the fluids. Buoyancy has an inverse relationship with density and the capillary pressure controls the entry pressure and relative permeability and also the effective sealing capacity. Gas and light to medium oil are lighter than water and this leads to overpressure (Fig. 3.3) (Osborne and Swarbrick 1997).

(vi) **Timing and rate**: Overpressure development depends upon the rate at which a system develops a non-permeable situation mostly in non-reservoir rock or presence of a shale zone. Swarbrick and Osborne (1998) stated that overpressure in a system involves dynamic processes where the first stage is the overpressure buildup phase during its generation and later with time there is overpressure

Table 3.1 Different overpressure generation mechanisms, tabulated in different categories. Modified after (Swarbrick and Osborne 1998)

Stress related	Fluid volume increase mechanisms	Fluid movement and buoyancy mechanisms
Disequilibrium compaction (vertical loading): Disequilibrium between fluid expulsion rate and sediment compaction due to fast burial. Fluids cannot be expelled and leads to overpressure	Temperature increase: Thermal expansion of water	Osmosis: Large contrasts in the formation fluids from dilute to saltier water across a semi permeable membrane
Lateral compressive stress: Incomplete dewatering on reduction of pore volume by horizontal stresses	Mineral transformation: smectite dehydration (Powers 1967) stated two to three pulses of water can significantly lead to overpressure in completely sealed sediments Gypsum to Anhydrite dehydration-loss of bound water Smectite-Illite transformation: Volume of water is released from smectite to illite	Hydraulic head: Recharge area pressure is exerted if it is overlain by seal
	Hydrocarbon generation: Volume increase during kerogen maturation during both oil and gas generation	

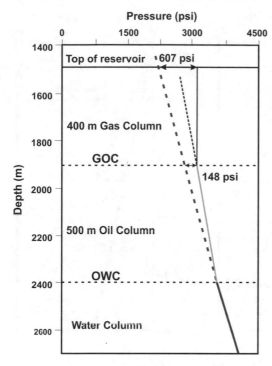

Fig. 3.3 Maximum pressure generation due to hydrocarbon buoyancy in North Sea (Swarbrick and Osborne 1998)

dissipation when there is leakage, as happens in shales. They also stated that the present scenario of an overpressured basin may depend on the present stress distribution of the area.

3.5 Pore Pressure Estimation Methods

Pore pressure estimation started becoming popular after the work of Hottman and Johnson's (1965), which was referred in the review by Zhang (2011). Hottman and Johnson (1965) studied the Oligocene and Miocene shales of Upper Texas and Louisiana Gulf coast. They plotted the sonic data against depth and found that porosity decreases with depth. This is a

general phenomenon of normally compacting sediments ("normal compaction trend"). Any data that deviates from this normal compaction trend line represents the abnormal zone with high fluid pressure.

Any attribute relating the change in pore pressure is used as pore pressure indicator. The estimation procedures involve two main approaches (Bowers 2001).

(i) Direct method: Two direct methods are commonly used, i.e., the Hottman and Johnson (1965) method and overlay method by Pennebaker (1968). These methods relate the deviation of the pore pressure indicator from the normal compaction trend line.

(ii) Effective stress methods: It works on the Terzaghi's principle where the difference between the total confining stress and the pore pressure controls the compaction of the sediments. It corresponds to the total stress carried out by mineral grains. These methods involve computational methods in three steps. (A) Effective stress (σ) calculation from pore pressure indicators, (B) bulk density is used to calculate the overburden stress, (C) pore fluid pressure (PP) is obtained from the algebraic difference between overburden stress and effective stress, as mentioned earlier in this section.

Late 1960s onwards all the pore pressure methods are based on effective stress methods and these are broadly divided into three categories (i) Vertical methods, (ii) Horizontal methods, and (iii) Other methods. We will very briefly mention below the first two methods.

(i) *Vertical Methods*: Foster and Whalen's (1966) Equivalent Depth method where the normal compaction trend value is used as the same as the pore pressure indicator value (Fig. 3.4).

Fig. 3.4 Vertical and horizontal methods of pore pressure estimation. Pore pressure indicator is used as the same value as the normal trend (point A) in vertical method of pore pressure estimation. Whereas, in horizontal method, the effective stress data is estimated from the normal trend (point B) (Bowers 2001)

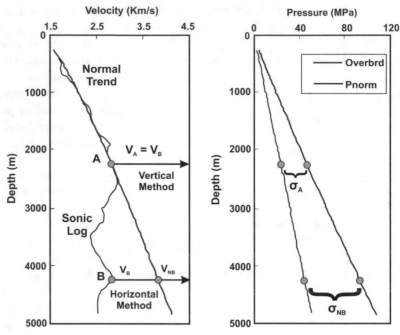

(ii) *Horizontal Methods*: Eaton's method (Eaton 1975) is one of the horizontal methods where the effective stress is calculated from the normal trend at the depth of interest (Fig. 3.4).

The classification of different pore pressure estimation methods is given as follows.

(a) Direct method by Hottman and Johnson and the Equivalent depth method

This methodology commonly deals with the departure of the pore pressure gradient from the normal velocity trend. This departure is empirically correlated and thus is not affected by pore pressure generation processes (Bowers 1995). The Hottman and Johnson (1965) analysis can be explained in terms of an X and a Y-axis (Fig. 3.4), where the X-axis represents the normal compaction trend, and the Y-axis the pore pressure gradient. This methodology is used for both sonic as well as resistivity data.

For resistivity data the equation is

$$X = Rn/R, \ Y = \text{Pore pressure gradient (psi/ft)} \tag{3.4}$$

For sonic data the equation becomes

$$X = \Delta t - \Delta t_n, $$
$$Y = \text{Pore pressure gradient (psi/ft)} \tag{3.5}$$

Subscript 'n'   value of normal trend.

Pore pressure development is not the same in every geological setting. Mathews and Kelly (1967) found that the chart for every area is different. The relationship of sonic and resistivity data in relation to the pore pressure gradient for different areas are shown in Figs. 3.5 and 3.6, respectively.

Eaton (1972) and Lane and Macpherson (1976) made a suggestion regarding the Hottman and Johnson method: the results or the output can

Fig. 3.5 Pore pressure
versus resistivity crossplot
(Owolabi et al. 1990)

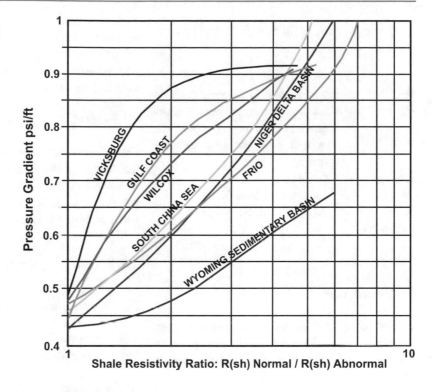

Fig. 3.6 Published pore
pressure crossplots for sonic
transit time (Owolabi et al.
1990)

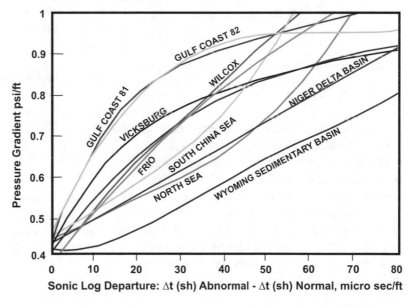

be more accurate if the data of overburden gradient is also used as an input. The H&J crossplots
vary with the overburden gradient and this is
particularly important in areas where the water

depth and salt thicknesses changes within few
km.

Another form of Hottman and Johnson data
was represented by Gardner et al. (1974):

$$P_f = \sigma_v - \left((\alpha_v - \beta)(A_1 - B_1 \ln \Delta t)^3 \right) \Big/ Z^2$$
$$(3.6)$$

P_f	formation fluid pressure (psi)
α_v	normal overburden stress gradient (psi/ft)
β	normal fluid pressure gradient (psi/ft)
Z	depth (ft)
Δt	sonic transit time (μs/ft)
A, B parameters	A1 = 82,776, B1 = 15,695.

(b) Pennebaker

Pennebaker (1968) predicted pore pressure from seismic velocity. He computed the pore pressure from X-Y crossplot similar to that of Hottman and Johnson.

$$X = \Delta t / \Delta t_n \text{ and Y is the pore pressure gradient} \tag{3.7}$$

The approximate equation of the crossplot is

$$Y = 1.017 - 0.531\, X^{-5.486} \tag{3.8}$$

Pennebaker's relations were based on the well data of Lousiana Gulf coast and Texas. He assumed that for similar rock type, the normal trend of interval transit time when plotted against depth is the same. With change in geological age and lithology, transit time would shift. Thus he proposed a single trend line usage for lithology worldwide. However, it was understood later that a single trend cannot be used for all types of lithologies.

(c) Vertical effective stress methods

This method assumes that in the case of both normally pressured and abnormally pressured scenarios, compaction takes place as a function of effective stress (Fig. 3.7; Bowers 1999).

Sometimes, normally pressured and over-pressured sediments may not follow the same effective stress relationship and Fig. 3.8 shows that the pore pressure can be under-estimated by the effective stress relationship.

(d) Equivalent depth method

The effective stress can be graphically solved by the equivalent depth method. From Fig. 3.6a the intersection of the vertical projection of the pore pressure and its normal trend, point A as indicated in Fig. 3.6 is termed as equivalent depth method.

$$P_B = OB_B - \sigma_A = OB_B - (OB_A - P_{NA}) \quad (3.9)$$

P_{NA}	hydrostatic normal pore pressure at point A
OB_B	Overburden pressure at point B
OB_A	Overburden pressure at point A.

The equivalent depth method was first used by Foster and Whalen (1966). Later Ham (1966) used this method with sonic, resistivity and density data.

(e) Mean stress equivalent depth

The modified version of equivalent depth was proposed by Traugot (1997). He defined:

$$\sigma_M = (\sigma + \sigma_h + \sigma_H)/3 \tag{3.10}$$

σ_M	mean effective stress
σ	vertical effective stress
σ_h	minimum horizontal stress
σ_H	maximum horizontal effective stress.

(f) Bellotti and Giacca's approach

Sonic wireline data is less sensitive to hole size variation, formation temperature and salinity of formation water, Fertl (1976) chose sonic data for pore pressure estimation. Out of the two sonic

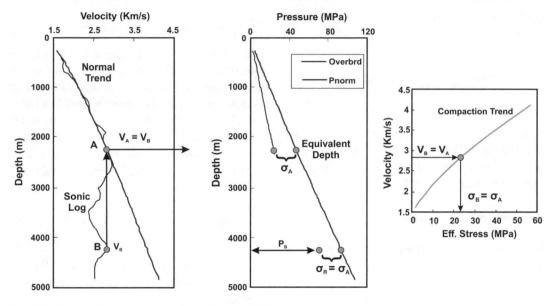

Fig. 3.7 Vertical effective stress methods (Bowers 2001)

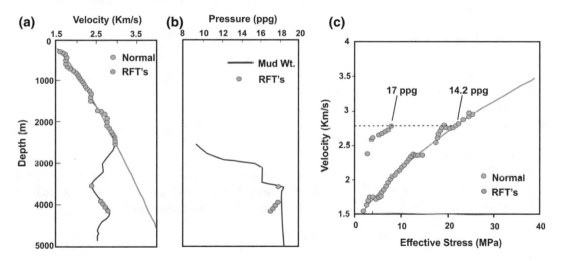

Fig. 3.8 Scenario where vertical effective stress method fails

based pore pressure estimation methods discussed in this part, one is by Bellotti and Giacca (1978) and the other one is by Hart et al. (1995).

$$V = V_{min} + (V_{max}\sigma/A\sigma + B) \quad (3.11)$$

V_{min}, minimum and maximum sonic
V_{max} velocity of the rock matrix, respectively

A, B additional calibration parameters
Σ vertical effective stress.

The density-velocity relation was also established by Bellotti and Giacca (1978) and derived

$$\Sigma = (V - V_{min})B/V_{max} - A(V - V_{min}) \quad (3.12)$$

the following equation for density from interval transit data.

$$P = \rho_{max} - 1.228(\rho_{max} - \rho_f)((\Delta t - \Delta t_{max})/\Delta t + \Delta t_f)$$
(3.13)

Zhang (2011) reviewed methods on pore pressure prediction. Different methods on pore pressure estimation methods from (interval) transit time were also included.

(g) **Eaton's (1975) method**

Sonic compressional transit time is used to predict pore pressure:

$$P_{pg} = OBG - \left(OBG - P_{ng}\right)(\Delta t_n/\Delta t)^3 \quad (3.14)$$

Δt_n interval transit time for the normal pressure in shales
Δt transit time of shale.

Zhang mentioned that Eq. 3.14 has a working limitation in those petroleum basins where the pore pressure generation is due to secondary mechanism (Detail in Chap. 4).

(h) **Bower's (1995) method**

A power relationship between sonic velocity and effective stress is presented:

$$V_p = V_{ml} + A\sigma^{Be} \quad (3.15)$$

V_p compressional wave velocity
V_{ml} mudline velocity
A, parameters which are calibrated with
B offset velocity and the effective stress data. "V_{ml}" is generally taken as 1520 ms^{-1}: equivalent to velocity near the sea floor. Using the equation $\sigma_e = \sigma_v - p$, the velocity dependent pore pressure can be derived from:

$$p = \sigma v - \left((V_p - V_{ml})/A\right)^{\left(\frac{1}{B}\right)} \quad (3.16)$$

The values A = 10–20 and B = 0.7–0.75 are taken from the Gulf of Mexico wells and the units for pressure measurements (p) and the effective stress σ_v are in psi. Velocities v_p and v_{ml} are in ft s^{-1}. The most common velocity measurement in wireline logs is sonic and it is in the form of transit time, the above Eq. 3.16 can be expressed in transit time by substituting the velocities v_p, v_{ml} as $10^6/\Delta t$ and $10^6/\Delta t_{ml}$, respectively.

$$p = \sigma v - \left(10^6\left(\frac{1}{\Delta t} - \frac{1}{\Delta t_{ml}}\right) \Big/ A\right)^{\left(\frac{1}{B}\right)} \quad (3.17)$$

The mudline compressional transit time is generally denoted by Δt_{ml} and the value is ~ 200 µs ft^{-1}.

To address the effect of unloading effect, Bowers (1995) introduced an empirical equation:

$$V_p = V_{ml} + A\left[\sigma_{max}(\sigma_e/\sigma_{max})^{1/U}\right]^B \quad (3.18)$$

U parameter representing uplift of the sediments.

$$\sigma_{max} = (V_{max} - V_{ml}/A)^{1/B} \quad (3.19)$$

σ_{max} and V_{max} are effective stress and velocity respectively and represent the starting point of unloading and the pore pressure for the unloading case can be estimated from the

$$P_{ulo} = \sigma_v - \left(V_p - V_{ml}/A\right)^{U/B}(\sigma_{max})^{1-U} \quad (3.20)$$

where P_{ulo} represents the pore pressure due to unloading behaviour (Zhang 2011).

(i) **Miller's method**

This method shows an inter-relation between velocity and effective stress and this relationship can be used to correlate the sonic/seismic transit time to formation pore pressure (Zhang 2011). The pore pressure can be obtained from:

$$P = \sigma v - \frac{\frac{1}{\lambda}\ln(V_m - V_{ml})}{(V_m - V_p)} \quad (3.21)$$

V_m sonic interval velocity with the shale matrix

V_m 14,000 to 16,000 ft s^{-1}

v_p compressional wave velocity at a particular depth and the rate of increase of velocity with effective stress is normally taken as 0.00025. Another important parameter that controls the occurrence of unloading is the "maximum velocity depth" (d_{max}). If d_{max} is less than the depth (Z) and there is no unloading (overpressure generated due to secondary mechanisms) then the pore pressure can be estimated from Eq. 3.21. If unloading happens where the $d_{max} > d$, then the pore pressure is:

$$P_{ulo} = \sigma v - \frac{1}{\lambda}\ln\left[a_m\left(1 - V_p - V_{ulo}/V_m - V_{ml}\right)\right] \quad (3.22)$$

a_m ratio of the loading and unloading velocities in the effective stress curves σ_{ul} where the values $a_m = 1.8$ and $a_m = V_p/V_{ulo}$ and V_{ulo} is the velocity where unloading begins. σ_{ul} represents the effective stress due to unloading of the sediment (Zhang 2011).

References

Bellotti P, Giacca D (1978) Pressure evaluation improves drilling performance. Oil Gas J

Biot MA (1941) General theory of three-dimensional consolidation. J Appl Phys 12:155164. https://doi.org/10.1063/1.1712886

Bowers GL (1995) Pore pressure estimation from velocity data; accounting for overpressure mechanisms besidesundercompaction. In: SPE drilling and completions, June

Bowers GL (1999) State of the art in pore pressure estimation. DEA-119 Report No 1

Bowers GL (2001) Determining an appropriate pore-pressure estimation strategy. In: Offshore technology conference 13042

Byerlee J (1993) Model for episodic flow of high-pressure water in fault zones before earthquakes. Geology 21:303–306

Dickinson G (1953) Geological aspects of abnormal reservoir pressures in Gulf Coast Louisiana. AAPG Bull 37:410–432

Donaldson EC, Chilingar GV, Robertson JO Jr, Serebryokov V (2002) Introduction to abnormally pressured formations. Developments of Petroleum Science, vol 50. Elsevier, Amsterdam, pp 1–19

Eaton BA (1972) The effect of overburden stress on geopressure prediction from well logs. J Pet Technol 24:929–934

Eaton BA (1975) The equation for geopressure prediction from well logs. In: Fall meeting of the society of petroleum engineers of AIME, Society of petroleum engineers

Engelder T (1993) Stress regimes in the lithosphere. Princeton University Press, Princeton

Fertl WH (1976) Abnormal formation pressures. Elsevier Scientific Publishing Co, New York, p 210

Foster JB, Whalen JE (1966) Estimation of formation pressures from electrical surveys Offshore Louisiana. J Petrol Technol 18(2):165–171

Gardner GHF, Gardner LW, Gregory AR (1974) Formation velocity and density—the diagnostic basis for stratigraphic traps. Geophysics 39:2085–2095

Ham HH (1966) A method of estimating formation pressures from Gulf Coast well logs. Gulf Coast Assoc Geol Soc Trans 16:185–197

Hart BS, Flemings PB, Deshpande A (1995) Porosity and pressure: role of compaction disequilibrium in the development of geopressures in a Gulf Coast Pleistocene basin. Geology 23:45–48

Hottman CE, Johnson RK (1965) Estimation of formation pressures from log-derived shale properties. J Pet Technol 17:717–722

Hunt JM (1990) Generation and migration of petroleum from abnormally pressured fluid compartments. AAPG Bull 74:1–12

Lane RA, Macpherson LA (1976) A review of geopressure evaluation from well logs-Louisiana Gulf Coast. J Petrol Technol 28(9):963–971

Law BE, Ulmishek GF, Slavin VI (1998) Abnormal pressures in hydrocarbon environments. In: Law BE, Ulmishek GF, SlavinVI (eds) Abnormal pressures in hydrocarbon environments. AAPG Memoir 70, pp 1–11

Louden LR (1972) Origin and maintenance of abnormal pressures. In: SPE 3843, 3rd symposium on abnormal subsurface pore pressure, Lousiana State University

Matthews WR, Kelly J (1967) How to predict formation pressure and fracture gradient from electric and sonic logs. Oil Gas J 20:92–106

Mukherjee S (2011) Estimating the viscosity of rock bodies-a comparison between the Hormuz-and the Namakdan Salt Domes in the Persian Gulf, and the Tso Morari Gneiss Dome in the Himalaya. J Indian Geophys Union 15:161–170

Mukherjee S, Kumar N (2018) A first-order model for temperature rise for uniform and differential compression of sediments in basins. Int J Earth Sci 107:2999–3004

Mukherjee S, Talbot CJ, Koyi HA (2010) Viscosity estimates of salt in the Hormuz and Namakdan salt diapirs, Persian Gulf. Geol Mag 147:497–507

Osborne MJ, Swarbrick RE (1997) Mechanisms for generating overpressure in sedimentary basins: a revaluation. Am Assoc Pet Geol 81:1023–1041

Owolabi OO, Okpobiri GA, Obomanu IA (1990) Prediction of abnormal pressures in the Niger Delta Basin using well logs. Paper No CIM/SPE 90-75

Pennebaker ES (1968) An engineering interpretation of seismic data. In: Fall meeting of the society of petroleum engineers of AIME. Society of Petroleum Engineers

Powers MC (1967) Fluid-release mechanism in compacting marine mud-rocks and their importance in oil exploration. AAPG Bull 51:1240–1245

Swarbrick RE, Osborne MJ (1998) Mechanisms that generate abnormal pressures: an overview. In: Law BE, Ulmishek GF, Slavin VI (eds) Abnormal pressures in hydrocarbon environments: AAPG Memoir 70, pp 13–34

Swarbrick RE, Osborne MJ, Yardley GS (2002) Comparison of overpressure magnitude resulting from the main generating mechanisms. In: Huffman AR, Bowers GL (eds) Pressure regimes in sedimentary basins and their prediction. AAPG Memoir 76, pp 1–12

Terzaghi K, Peck RB, Mesri G (1996) Soil mechanics in engineering practice, 3rd edn. Wiley, Hoboken

Traugott M (1997) Pore/fracture pressure determinations in deep water. Deepwater technology, supplement to August. World Oil 218, pp 68–70

Zhang J (2011) Pore pressure prediction from well logs: methods, modifications, and new approaches. Earth-Sci Rev 108:50–63

Detection of Abnormal Pressures from Well Logs

4

Abstract

Continuous attributes through depth are obtained using wireline logs and logging while drilling. A number of well logging techniques enables detection of overpressure zones. How porosity link with pore pressure is the main key to detect the abnormal pressure. Abnormal pressure, i.e., overpressure and under pressure, can be quantified by noting how much the depth-wise log-data for a rock type varies from that of a shale. Sonic logs can better detect abnormal pressure zones than the neutron and the density logs. Effective stress reduction opens connecting pores easier than the storage pores. This chapter explains pore pressure mechanisms using cross plots of wireline logs.

4.1 Introduction

Wireline logs and logging while drilling (LWD) are the techniques that provide continuous attributes through depth (Bigelow 1994). Detection of abnormal pressure zones in low permeability rocks such as shales started in 1960s using different well logging techniques

The original version of this chapter was revised: The figure 4.10 was inadvertently published without proper permissions. The correction to this chapter is available at https://doi.org/10.1007/978-3-030-13442-6_8

(Hermanrud et al. 1998). Well logs such as resistivity, sonic, bulk density and neutron responds to the delineation of abnormal pressure zones (Ramdhan et al. 2011). The main assumption in detecting an abnormal pressure zone from well logs relates to the interrelationship between porosity and pore pressure (Hermanrud et al. 1998). The continuous log data plotted versus depth and the trend lines developed from normally compacted shales are referred as the normal compaction trend (Bigelow 1994). As mentioned in Chap. 3, deviations from this normal trend line represent the abnormal zone and the degree of abnormality depends on the rate of departure from the hydrostatic pressure (Dickinson 1953).

Compaction disequilibrium and unloading are the two overpressure generating processes (Bowers 2001; Ramdhan et al. 2011) and these mechanisms can be easily distinguished using wireline logs (e.g., Ramdhan et al. 2011). Abnormal pressure scenarios are masked by neutron and density logs whereas the sonic resistivity log responses are sensitive to abnormal pressures (e.g., Ramdhan et al. 2011). Neutron- and density logs measure the bulk properties (such as matrix density) whereas sonic logs measure transport properties (e.g., Ramdhan et al. 2011). Bowers and Katsube (2002) classified the shale pore structures into storage and connecting pores. Based on laboratory results, they concluded that the connecting pores opens easier than the storage pores because of effective stress reduction during overpressure detection.

© Springer Nature Switzerland AG 2020, corrected publication 2020
T. Dasgupta and S. Mukherjee, *Sediment Compaction and Applications in Petroleum Geoscience*,
Advances in Oil and Gas Exploration & Production, https://doi.org/10.1007/978-3-030-13442-6_4

As a result of overpressure due to compaction disequilibrium can be identified through resistivity, density, neutron and sonic logs (Bowers and Katsube 2002) whereas the overpressure generated by secondary mechanisms e.g., hydrocarbon generation, clay diagenesis etc. can be detected by sonic and resistivity logs (Bowers and Katsube 2002). Density versus sonic transit time (Bowers 2001; Dutta 2002; Katahara 2006; Ramdhan et al. 2011) and density versus resistivity crossplots (Ramdhan et al. 2011) detect overpressured zones caused by unloading. This chapter gives an overview to the readers about the identification of different pore pressure causing mechanisms through cross plots of wireline logs.

4.2 Wireline Log Responses

Few pore pressure estimation techniques, especially the "equivalent depth method", are used worldwide (Hermanrud et al. 1998). These methods are developed on the responses from homogeneous shales at the U.S. Gulf Coast and are based on the empirical relationship between shale porosity and burial depth. Recent techniques use Terzaghi's relationship, which mainly deals with effective stress (Hermanrud et al. 1998) (Chap. 3). Information on effective stress versus sonic velocity or porosity can distinguish between overpressure generation mechanisms: compaction disequilibrium and fluid expansion (Hermanrud et al. 1998).

Hermanrud et al. (1998) studied the wireline log responses of the intra reservoir shales (offshore Norway; Fig. 4.1). 28 well data were studied out of which 6 wells are from highly pressured intervals and 22 from normally or moderately pressured intervals.

Shallow wells were also studied to establish the normal compaction trend line by plotting porosity data from the Not Formation versus depth. Shallow wells are taken for normal compaction trend studies because by studying the surface porosity data one would be able to know the rate of porosity decay with increase in overburden. Core-log integration was done for the Garn Formation. The main objective of

Hermanrud et al. (1998) was to identify the log response of the Not Formation shale present in all the normal, moderate and high pressure regimes. These authors calculated the average porosities of the shales from density, neutron, resistivity and sonic logs. Figures 4.1 and 4.2 shows the typical log response of the Not- and the Ror Formation.

Porosity from density log is calculated from the following relationship:

$$\varnothing_{RHOB} = (\rho_{ma} - \rho_b)/(\rho_{ma} - \rho_{fl}) \qquad (4.1)$$

Here ρ_{ma}: matrix density, taken as 2.72 g cm^{-3}, ρ_b: bulk density measured by log, ρ_{fl}: fluid density, taken as g cm^{-3}.

The porosity calculations from sonic data involves inputs such as interval transit time of the matrix (Δt_{ma} = 68.8 μs ft^{-1}, 226 μsm^{-1}), the fluid transit time is (Δt_{fl} = 189 μs ft^{-1}, 620 μs m^{-1}). Δt: average interval transit time from log.

$$\varnothing_{DT} = (\Delta t - \Delta t_{ma})/(\Delta t_{fl} - \Delta t_{ma}) \qquad (4.2)$$

Porosity from resistivity log ($\varnothing_{RILD}$) was calculated from Archie's law (1952) for water bearing sands:

$$\varnothing_{RILD} = \{a(R_w/R_t)\}/(S^n_w) \qquad (4.3)$$

The formation water resistivity R_w = 0.0294 m, saturation exponent n = 2 and cementation exponent m = 2 are taken to calculate porosity. The average true resistivity of the formation R_t = 6 Ωm is used. The neutron porosities range 24–28 and 8–31% for the Not- and the Ror Formation, respectively. For the sandstone-bearing Garn Formation, ρ_{ma} = 2.65 g cm^{-3} and Δt_{ma} = 55 μs ft^{-1} were taken as input for porosity calculated from density- and sonic logs respectively.

Hermanrud et al. (1998) attempted to find differences in porosity amongst the three pressured regimes. The actual magnitude of the porosities was unimportant but only the relative values between the pressure regimes were considered. Density, neutron, resistivity and sonic log derived average porosities were plotted against depth for each of the three pressure regimes (Fig. 4.3).

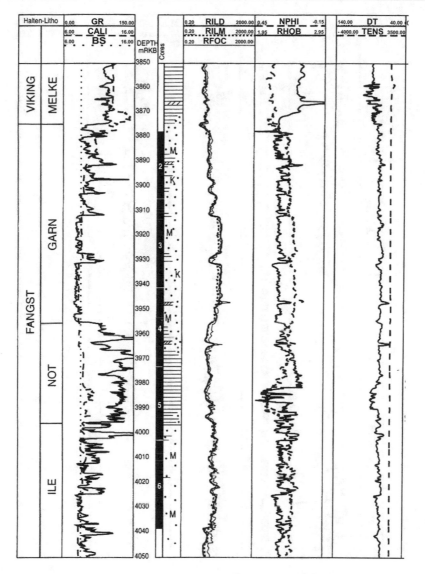

Fig. 4.1 Log pattern of Not formation along with helium calculated porosities. Some parts of the Not formation are affected by poor hole conditions (3968–3993 m). Modified after Hermanrud et al. (1998)

Hermanrud et al. (1998) observed that the density and neutron derived average porosities in the high pressured regimes were similar to those encountered in the normal pressured regimes. Whereas the resistivity and sonic derived porosities are higher in the overpressured zones with pore pressure up to 1.8 g cm^{-3} (15 ppg equivalent mud weight) than the normal ones. Similar observations were made by the authors for average porosities derived from density, neutron, resistivity and sonic logs in the Ror

Formation (Fig. 4.4). Table 4.1 presents porosity derived from each log.

Sonic- and resistivity logs measure the transport properties whereas neutron- and density logs measure bulk properties (Hermanrud et al. 1998; Bowers and Katsube 2002; Ramdhan et al. 2011). Hermanrud et al. (1998) suggested that textural changes are sensitive due to overpressure whereas the resistivity logs were reacting to increased fluid conductivity. Sonic velocity reduced since the intergranular stresses

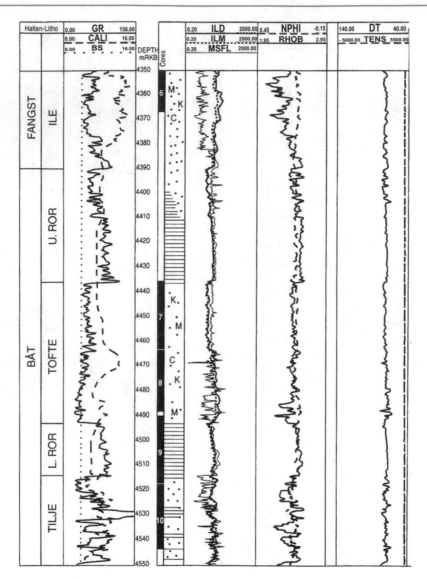

Fig. 4.2 Log pattern of Ror formation along with helium calculated porosities. Some average porosities were calculated for all the shaly formations in the Upper and lower Ror formations. Modified after Hermanrud et al. (1998)

decreased. The actual cause of increase in over-pressure remained indeterminate (Hermanrud et al. 1998).

Compaction disequilibrium due to rapid burial of the sediments primarily generates overpressure (e.g., Ramdhan et al. 2011). Reducing stress by unloading is the secondary mechanism that generate overpressure (Yardley and Swarbrick 2000; Bowers 2001; Bowers and Katsube 2002; Lahann 2002; Swarbrick et al. 2002; Ramdhan et al. 2011).

Bowers and Katsube (2002) in their review on sensitivity of wireline log responses mentioned the observations by Hermanrud et al. (1998). To identify the exact mechanism for pore pressure generations, Toksoz et al. (1976) and Cheng and Toksoz (1979) conducted oblate pore modelling to take care of pressure sensitivity. The pores were classified according to aspect ratio (α): (i) elastically rigid: $\alpha > 0.1$, e.g., vugs and intergranular pores; (ii) elastically flexible,

Fig. 4.3 Average porosities calculated in Not formation from **a** Sonic (DT) log, **b** Density (Rhob), **c** Resistivity (Rt) and **d** neutron log. Each of the pressure regimes are indicated by different symbols. Modified after Hermanrud et al. (1998)

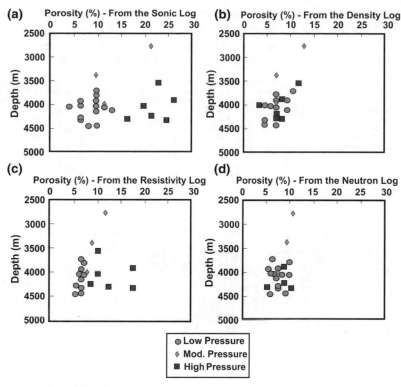

Fig. 4.4 Average porosities calculated in Ror Formation from **a** Sonic (DT) log, **b** Density (Rhob), **c** Resistivity (Rt) and **d** neutron log. Each of the pressure regimes are indicated by different symbols. Modified after Hermanrud et al. (1998)

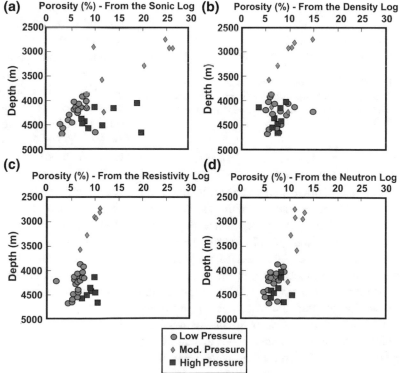

Table 4.1 Average porosities calculated in the Ror Formation from (a) Sonic (DT) log, (b) Density (Rhob), (c) Resistivity (Rt), and (d) neutron log (Hermanrud et al. 1998)

	Average log-derived porosity: Not formation				Average log-derived porosity: upper Ror formation				Average log-derived porosity: lower Ror formation			
Log	LP wells	N	OP wells	N	LP wells	N	OP wells	N	LP wells	N	OP wells	N
Sonic (DT)	8.6	14	21.8	6	5.8	14	10.8	5	6	7	12.1	4
Density (RHOB)	7.3	13	7.6	6	7.7	14	6.9	4	7.5	8	7.3	4
Neutron (NPHI)	7.1	13	7.8	4	6.6	13	8.3	4	6.7	8	7.5	4
Resistivity (RILD)	6.3	14	12.5	6	6.6	14	8.9	4	5.3	8	9.2	3

Basic Pore Types

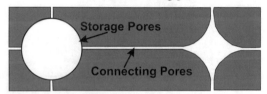

Storage Pores
High aspect ratios
Mechanically stiff
'Nodes' along the pore network

Connecting Pores
Low aspect ratios
Mechanically flexible
Control transport properties during rebound

Fig. 4.5 Pore structure model of a rock, constituting storage pores with high aspect ratio and connecting pores with low aspect ratio. Modified after Bowers and Katsube (2002)

$\alpha = 0.001 - 0.1$; and (iii) collapsing pores: $\alpha < 0.001$ (Bowers and Katsube 2002). The high aspect ratio storage pores connect to the lower aspect ratio pores and constitute the rock pore structures (Fig. 4.5). These connecting pores control the fluid transport. In lab experiments, the response of connecting pores to the laboratory stress changes exceeds that of the storage pores.

Bowers and Katsube (2002) proposed the sensitivity of unloading i.e. due to secondary generation of overpressure to the ratio of the volume of the connecting pores to that of the storage pores. The connecting pores being flexible can open by the unloading effect, but the stiff storage pores remain unaffected (Bowers and Katsube 2002; Ramdhan et al. 2011). Thus, the transport properties such as sonic velocity and the electrical conductivity are affected by the opening of the connecting pores due to overpressure conditions, whereas the bulk rock properties as well as the neutron log remain unaffected (Hermanrud et al. 1998; Bowers and Katsube 2002; Ramdhan et al. 2011). Due to unloading process of overpressure generation, sonic transit time and resistivity shows a reversal trend when plotted against depth (Fig. 4.6; Hermanrud et al. 1998; Bowers and Katsube 2002; Ramdhan et al. 2011).

The common pressure generating mechanism seen in the U.S Gulf Coast wells is due to compaction disequilibrium but the log signatures in some wells of the Gulf of Mexico depicted the influence of unloading due to overpressure generation (Fig. 4.7; Bowers and Katsube 2002; Katahara 2006).

Certain cross-plots such as density versus sonic velocity is used to detect unloading (Fig. 4.8) (Bowers 2001; Dutta 2002; Katahara 2006; Ramdhan et al. 2011). The data falling on the compaction curve represents the normal to mild overpressure due to compaction disequilibrium whereas the data showing velocity reversal relates to unloading.

Some recent methods such as effective stress versus density and sonic relationships also delineate the loading limb to the unloading limb (Bowers 2001; Hoesni 2004; Tingay et al. 2011) (Fig. 4.9).

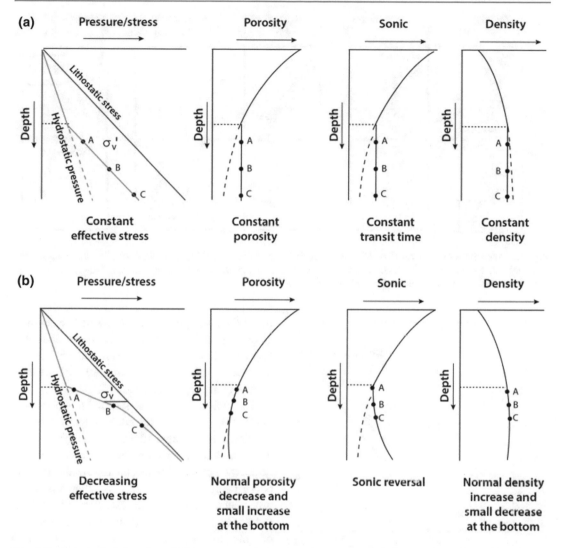

Fig. 4.6 Cartoon depicting the wireline responses in **a** overpressure generated due to compaction disequilibrium process, **b** overpressure generated due to unloading process (Ramdhan et al. 2011)

The departure from the compaction trend can be because of two reasons, one as the rock property changes and the other by overpressure. Clay diagenesis from smectite to illite conversion reverses velocity as per Fig. 4.10 (Dutta 1987; Bowers 2001; Hoesni 2004; Ramdhan et al. 2011). Clay diagenes is reduces effective stress during smectite-illite conversion as the stress transfers from smectite grains to pore water during conversion. As stated in (Chap. 3) this phenomenon results in pore pressure more than the overburden stress.

Hoesni (2004) studied the wells in and around Malay basin (offshore Malaysia) to infer the causal mechanisms of overpressure (Fig. 4.11). Velocity–density cross-plots were also generated in order to support the interpretations regarding the pressure generation (Table 4.2). Later Tingay et al. (2009, 2013) made detailed studies by integrating wireline formation tester data. Such data were recorded in four formations 2A-D the Malay basin. Sonic velocity-effective stress relationship was used to distinguish data representing the unloading behaviour and loading pattern.

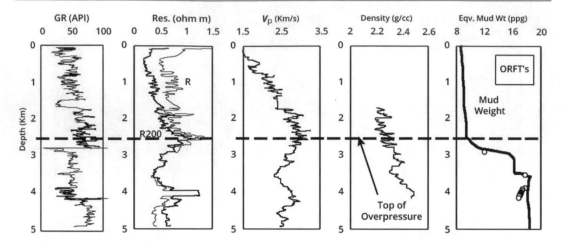

Fig. 4.7 Well from Gulf of Mexico showing the typical response due to unloading. At the onset of overpressure zone the extent of response shown by sonic velocity and resistivity log is more than that of the density log. Modified after Bowers and Katsube (2002)

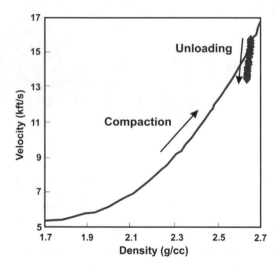

Fig. 4.8 Velocity versus density crossplot for identifying unloading behaviour. Modified after Bowers (2001)

Sonic data was taken as proxy for porosity and the vertical effective stress was calculated from available density logs and pore pressure determined from drill stem testing (DST) and wireline formation testers. Empirically calibrated Nafe-Drake's relationship was used to integrate the density from surface up to top of the available logs and there was integration of velocities calibrated with checkshot data (Tingay et al. 2003).

Sonic data was preferred over porosity tools such as density and neutron because the sonic tool is less affected by borehole conditions and can be integrated with seismic-based pore pressure estimation (Tingay et al. 2009).

The wireline foreline formation testers and DSTs were plotted in the sonic velocity plot and the calculated vertical effective stress (Fig. 4.12). 725 wireline formation testers represent the normally pressured data and follow the loading curve (Tingay et al. 2013). Wireline formation tester data of 21 wells representing mild to high (>14 MPa km^{-1}) overpressured formations are off the loading curve (Tingay et al. 2013) (Fig. 4.12). Some data fall around the scatter of the loading curve (Tingay et al. 2009). Some data represent the normal loading curve. The data off the loading curve represent overpressure due to fluid expansion mechanism, load transfer and due to vertical transfer (Tingay et al. 2013).

Three formations 2A-2B-2C were studied separately and an effective stress versus sonic velocity plot was made for each of these formations. The normally pressured and mild overpressured (11.5–14 MPa km^{-1}) data of each formation follow the loading curve whereas the overpressured data follow a trend off the loading curve (Fig. 4.13; Tingay et al. 2013).

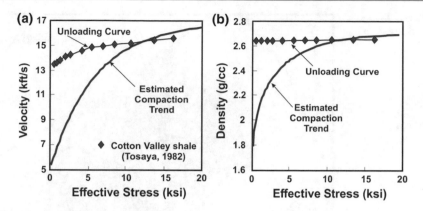

Fig. 4.9 Detection of unloading behaviour **a** Velocity versus effective stress, velocity reversal along with reduction in effective stress. **b** Density versus effective stress, effective stress reduction but density remains constant (Bowers 2001)

Fig. 4.10 Overpressure trends possibly due to unloading (after Hoesni 2004). **a** Fluid pressure versus depth. **b** Shale porosity versus depth. **c** Shale porosity versus effective stress. **d** Bulk density versus velocity

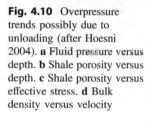

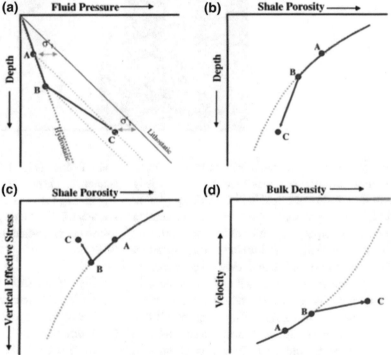

Tingay et al. (2013) concluded that the overpressured formations in the N Malay basin mostly follow the trend off the loading curve and fluid expansion/transfer mechanism causes overpressure. Tingay et al. (2013) plotted some high mobility data (>10 md/cp) from wireline formation tester measurements (Fig. 4.14).

A relation between velocity and effective stress cannot distinguish fluid expansion and vertical transfer mechanisms (Tingay et al. 2007, 2013). Velocity versus density crossplot differentiate them (Fig. 4.15). (Hoesni 2004; Lahann and Swarbrick 2011; O'Conner et al. 2011; Tingay et al. 2013).

Sediments deposited and compacting with subsequent burial represent normal pressure conditions and follow the loading curve in sonic-velocity cross-plot. Sequences in which

Fig. 4.11 Map of Malay
basin with well locations
(Hoesni 2004)

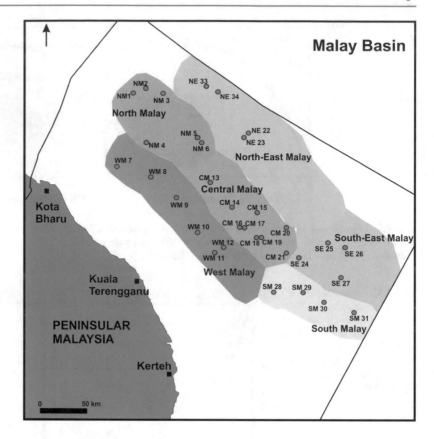

overpressure is generated by compaction disequilibrium processes consists of similar porosity
and thus follow the loading curve (Tingay et al.
2013). Kerogen maturation and gas generation
increase overpressure and this reduce drastically
the sonic velocity with little/no density variation
(Hoesni 2004). Whereas, clay diagenesis/load
transfer increases density with minor variation in
sonic velocity (Lahann and Swarbrick 2011;
O'Conner et al. 2011; Tingay et al. 2013).
Combination of different mechanisms such as
gas generation and clay diagenesis changes in
velocity with increasing pressure. The above
mentioned crossplots were generated to understand overpressure mechanism in the northern
Malay basin (Fig. 4.16; Tingay et al. 2013).

The normally pressured sequences present
above and below the mild to overpressured (pore
pressure < 17 MPa km^{-1}) wells of northern
Malay basin follow the typical loading curve. In

wells B & D (Fig. 4.16), the normally pressured
sequences show linear trend in the sonic—density cross-plot with 2A and 2C as overpressured
sequences. The overpressured zones show the
following characteristics (Fig. 4.16; Tingay et al.
2013):

(a) Significant drop in sonic velocity whereas
    the density data varies insignificantly within
    the overpressured reservoir
(b) From the middle of the formations 2A to 2B,
    there is gradual increase in the sonic as well
    as density value.
(c) Finally pore pressure gradient falls near the
    contact of 2A and the normally pressured
    sequences, sonic velocity increases without
    much variation in density.

Highly overpressured sequences in the wells
(E & D) with pressure gradient (>17Mpa km^{-1})

Table 4.2 Well list with dominant pressure mechanisms (denoted by symbols) on the basis of pore pressure profile and velocity-density crossplot

Well name	Pressure profile	Velocity-density crossplots
NM-1	D + C/U	D + C
NM-3	D + C/U	D + C
NM-5	D	D + C
NM-6	D	D + C
WM-7	N	N
WM-8	D + C/U	D + C?
WM-10	D	D + C
WM-11		C + D?
CM-15	D	D
CM-16		D
CM-17	D + U?	D
CM-18		D + C?
CM-19		N
NE-22		X
NE-23		X
NE-33		X
NE-34		X
SE-24		D
SE-25	D	D
SE-26		C
SE-27		D
SM-28	D	D + C
SM-29	D	D
SM-30	C + D	C + D
SM-31	C + D	C + D

D disequilibrium compaction, *U* unloading, *C* chemical compaction, *N* normal compaction, *X* reason unknown (Hoesni 2004)

are present towards the basin center (Fig. 4.17). These wells do not show any signature of pressure reduction at the contact of 2A and normally pressured sequences as seen in the other part of the basin. The overpressured wells show a sharp drop in sonic velocity, with slight reduction in density and then there is a sharp increase in the sonic velocity and the density remains constant in the Formation 2A and 2B (Fig. 4.17). Even if the pore pressure shows an increasing or decreasing trend with respect to hydrostatic pressure, the sonic velocity shows huge variation with respect to density data in the northern Malay Basin sequences. This sonic–density signature in overpressured zones depicts the impact of kerogen maturation and gas generation. In the northern Malay Basin overpressure is also generated due to clay diagenesis or load transfer as shown by the response of sonic density cross-plot where both sonic velocity and density show a returning trend towards the loading curve. These signatures correspond to the lowering of pressure gradient at the hydrostatic equilibrium. High overpressure zones display sharp reduction in sonic and density which return to the loading curve. This signifies that the high overpressure generation is due to compaction disequilibrium along with gas generation.

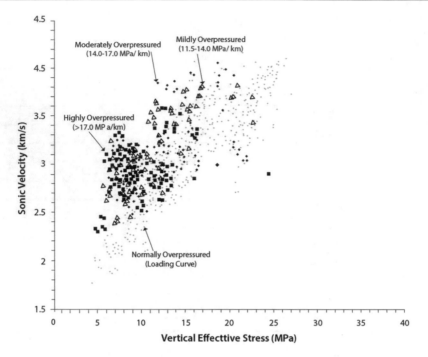

Fig. 4.12 Sonic velocity-effective stress plot showing the normally pressured data fallowing loading curve whereas Mildly overpressured (11.5–14 MPa km^{-1}), Moderately Overpressured (14–17 MPa km^{-1}) & Highly Overpressured (>17 MPa km^{-1}) data falling off the loading curve. Overpressure in the formation following the unloading curve is because of the fluid expansion, load transfer or vertical transfer mechanism. Most of the data representing unloading curve (Tingay et al. 2013)

The Brunei Basin (of Brunei) shows the development of complex overpressure (Hoesni 2004; Tingay et al. 2005, 2009). The Baram delta province, within this basin, is composed of three deltaic systems (Tingay et al. 2009), i.e., the Meligan delta (Early Miocene), the Champion delta (Early to Late Miocene) and the Baram delta (Late Miocene to present).

Shale diapirs, mud volcano sand shale dykes exist in the Baram delta province and numerous faults and dykes were encountered during drilling that resulted in kicks and blow-outs (Tingay et al. 2005, 2009). The pore pressure gradient in these wells exceeds 11.5 MPa km^{-1}. According to Tingay et al. (2003, 2009) overpressure in Brunei is mainly due to pro-delta shale, near gas chimneys, shale diapirs and in uplifted areas with high geothermal gradient. The overlying deltaic sediments also developed significant overpressure. The sonic data was used as a proxy for porosity. Porosity versus effective stress data was used to distinguish overpressure produced by compaction disequilibrium and that by fluid expansion. Vertical effective stress was calculated for 1400 reservoir formation testers (RFTs) from 90 wells in 31 fields and was plotted in velocity versus effective stress plot. The loading curve of the velocity–effective stress plot was calculated from the 825 normally pressured pore pressure RFT data of 11.5 MPa km^{-1}. Wells in which overpressure is due to compaction disequilibrium fall on the loading curve. Overpressure generated by the fluid expansion mechanism fall off the loading curve (Fig. 4.17). Along the compaction disequilibrium mechanism, overpressure is vertically transferred in the deltaic sediments from the pro-delta shale (Tingay et al. 2007, 2009).

Highly overpressured shales are encountered while drilling wells in Tertiary basin like the Niger Delta which lead to frequent losses, cavings, stuck pipe, etc. (Nwozor et al. 2013). The top of overpressure varies: 1370 m true vertical depth below

Fig. 4.13 Sonic
velocity-effective stress plot
of the three formations 2A, 2b
and 2C. Dark grey dots with
pressure >11.5 MPa km^{-1}
(0.51 psi ft^{-1}) represent the
unloading curve and the
normally pressured
(hydrostatic pressure) wireline
formation tester data represent
the loading curve (Tingay
et al. 2013)

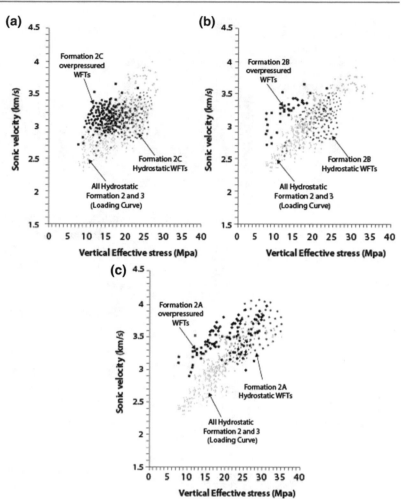

subsea (tvdss) to 4270 mtvdss. Like other young
Tertiary basins such as the Gulf of Mexico, Nile
delta and the Baram basins, the Niger delta also
shows high sedimentation rate and deposition of
thick shales. In such areas, compaction disequilibrium is considered as one of the dominant
pressure generating mechanisms. Post-sediment
burial secondary mechanisms generates pressure
as discussed earlier. To identify the dominant
pressure generating mechanisms, vertical effective
stress versus density, vertical effective stress versus velocity, and velocity versus density cross-
plots have been studied (Nwozor et al. 2013).
Amongst the data from the four wells A, B, C and
D, well D is hydrostatically normally pressured
and the loading curve is modelled from the well
data D (Fig. 4.18). Nwozor et al. (2013) concluded

that the deviation of data from the loading curve
signifies that the fluid expansion mechanism was
responsible for overpressure generation. This
matches general observations by Chopra and
Huffman (2006).

Vertical effective stress versus sonic velocity
cross-plot was made by Nwozor et al. (2013).
The normal compaction trend is defined by the
non-reduction of the effective stress, whereas
certain data especially those of wells A and B
reduce sharply in vertical effective stress and
velocity reversal. Thus, the data which are falling
off the loading curve signifies overpressure generation due to unloading (Fig. 4.19).

Velocity versus vertical effective stress and
velocity versus density cross-plot signify
unloading mechanism apart from compaction

Fig. 4.14 Sonic velocity-effective stress plot of 49 data with excellent mobility (>10 mD/cp). Low mobility formation show inaccurate measurements (Tingay et al. 2013)

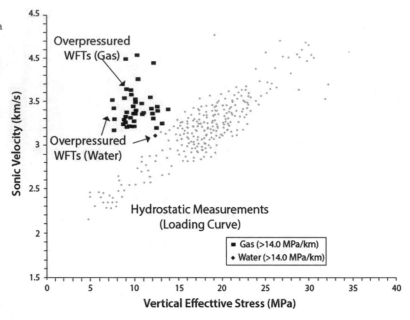

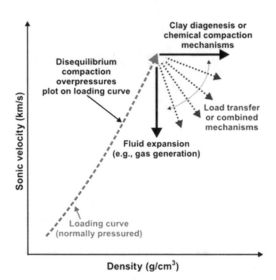

Fig. 4.15 Sonic velocity versus density to distinguish overpressure generated by fluid expansion mechanism (Adapted from Hoesni 2004; O'Conner et al. 2011; Tingay et al. 2013). Data of the formation in which overpressure is generated by compaction disequilibrium follow the loading curve. Mechanisms related to clay diagenesis and load transfer shows drastic increase in the density and little change in the sonic velocity. Overpressure generation due to gas generation shows severe drop in sonic velocity with little or no increase in density

disequilibrium as one of the dominant mechanism of overpressure generation.

An overpressure distribution study was done by John et al. (2014) in the Mahanadi Basin, north east coast block (NEC) of India. Pressure stratigraphy was made using an integrated approach using offset well data and seismic velocities. The Miocene level shows variation in pressure from northern to southern wells (Fig. 4.20). Pore pressure ranges between 13.3 and 15.7 MPa km^{-1} within shelfal and basinal areas, respectively. Overpressured zones (19.7 MPa km^{-1}) were encountered during drilling of wells towards the northern part whereas the wells towards the southern part show moderate to high pore pressure: 13.7 MPa km^{-1}. In order to identify the dominant overpressure generating mechanism, velocity versus effective stress and sonic velocity versus density cross-plots were made (John et al. 2014). Effective stress versus sonic velocity cross-plot (Fig. 4.20) for the data of northern part of the NE coast of basin shows that the pore pressure exceeding 11.5 MPa km^{-1} fall off the loading curve. The southern part of the basin with pore pressure

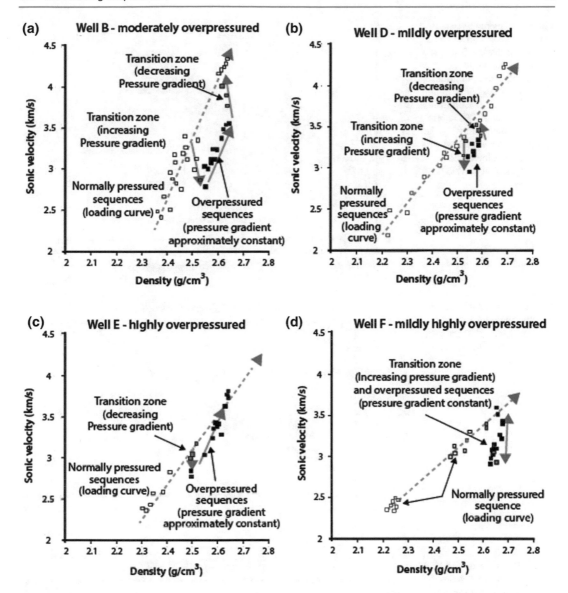

Fig. 4.16 Cross plot between Sonic velocity versus density for four wells in Malay basin. The normally pressured sediments are represented by white squares and they follow the loading curve. The loading curve are the data of normally pressured formations and are shown by white squares. As discussed in the text, the part of the sequences within the overpressure transition zone is represented by gray squares, the center of the overpressured zone is represented by black squares (After Tingay et al. 2013)

exceeding 11.5 MPa km^{-1} follows the loading curve.

Thus, the overpressure in northern part of the Mahanadi Basin is by unloading whereas undercompaction played an important role in overpressure generation in the southern part. John et al. (2014) also plotted the sonic velocity versus density for the northern as well as southern part of the basin (Fig. 4.21).

In the sonic velocity versus density cross-plot for the northern part of the Mahanadi Basin, the pore pressure of the wireline formation tester data >11.5 MPa km^{-1} shows drastic reduction of sonic velocity whereas the density values are

Fig. 4.17 Porosity (sonic
velocity) versus effective
stress plot distinguishing the
overpressure due to
compaction disequilibrium
(black dots) and fluid
expansion mechanism (ash
colored squares). The normal
pressured data following the
loading curve is defined by
small grey colored dots
(after Tingay et al. 2009)

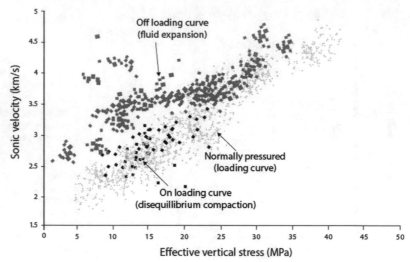

Fig. 4.18 Velocity versus
density crossplot from the
onshore wells of Niger delt.
Adapted after Nwozor et al.
(2013)

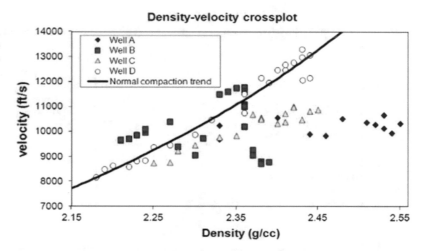

Fig. 4.19 Velocity versus
vertical effective stress
crossplot from the onshore
wells of Niger delt
(Nwozor et al. 2013)

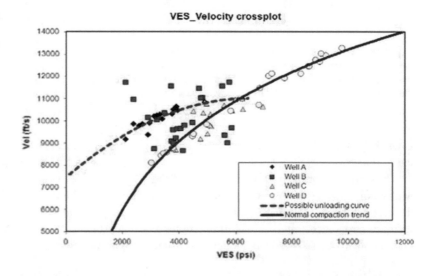

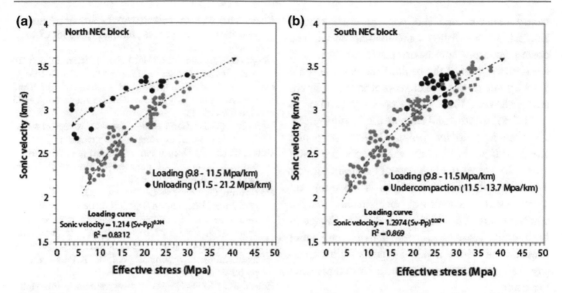

Fig. 4.20 a Sonic velocity versus effective stress cross-plot for the wells in the northern part. 170 Wireline formation tester data were used to calculate the vertical effective stress. Pore pressure data >11.5 MPa km^{-1} fall off the loading curve and delineate the unloading zones,

b Pore pressure from 140 wireline formation tester data were plotted in sonic velocity versus effective stress plot for southern NEC block, all the pore pressure >11.5 MPa km^{-1} follow the loading curve and signifies overpressure due to undercompaction (John et al. 2014)

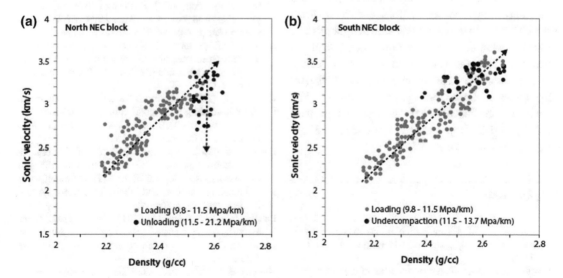

Fig. 4.21 a Sonic velocity versus density crossplot for the wells in the northern part. 170 Wireline formation tester data were used for the study. Pore pressure data >11.5 MPa km^{-1} fall off the loading curve and delineate the unloading zones representing sharp reduction in sonic velocity and constant density. **b** Pore pressure from 140

wireline formation tester data were plotted in sonic velocity versus density plot for southern NEC block, all the pore pressure >11.5 MPa km^{-1} follow the loading curve and signifies overpressure due to undercompaction (John et al. 2014)

nearly constant and the pore pressure around 11.5 MPa km^{-1} follows the loading curve. The onset of overpressure is from 2800 to 2900 m in the northern part of the basin. Whereas, the sonic velocity versus density cross-plot in the southern part shows that the pore pressure is ~11.5 MPa km^{-1} and follows the loading curve. Thus sonic velocity versus density cross-plot along with sonic velocity versus effective stress cross-plot across the NEcoast of the basin suggest that compaction disequilibrium is the dominant overpressure generating mechanism in the southern part of the basin. The distinct change in the sonic velocity versus density cross-plot towards the northern part suggests fluid expansion to be the dominant process of overpressure generation.

John et al. (2017) deduced a methodology to establish a relationship between overpressure and the Vp/Vs ratios. Overpressure caused by undercompaction can easily be deduced through the effective stress versus velocity relationship whereas the unloading mechanism does not show direct relationships. In the case study shown by John et al. (2017), chemical compaction mechanism is deduced from the sonic velocity versus density relationship. The shear velocity being unaffected by the effect of fluid, the Vp/Vs relationship thus enhances the effect of fluid in the system. John et al. (2017) used the drop in the Vp/Vs trend to delineate the overpressure zone.

References

Archie GE (1952) Classification of carbonate reservoir rocks and petrophysical considerations. AAPG Bull 36:278–298

Bigelow EL (1994) Global occurrences of abnormal pressures. In: Fertl WH, Chapman RE, Hotz RE (eds) Studies in abnormal pressures. Development in petroleum science 38. Elsevier Science, Amsterdam, pp 1–17

Bowers GL (2001) Determining an appropriate pore-pressure estimation strategy. In: Offshore technology conference, vol 13042

Bowers GL, Katsube TJ (2002) The role of shale pore structure on the sensitivity of wire-line logs to overpressure. In: Huffman AR, Bowers GL (eds) Pressure regimes in sedimentary basins and their prediction, vol 76. AAPG Memoir, pp 43–60

Cheng HC, Toksoz MN (1979) Inversion of seismic velocities for the pore aspect ratio spectrum of rock. J Geophys Res 84:7533–7543

Chopra S, Huffman AR (2006) Velocity determination for pore pressure prediction. Lead Edge 25:1502–1515

Dickinson G (1953) Geological aspects of abnormal reservoir pressures in Gulf Coast Louisiana. AAPG Bull 37:410–432

Dutta NC (1987) Geopressure, geophysics reprint series No. 7, Society of exploration geophysicists

Dutta NC (2002) Deep water geohazard prediction using prestack inversion of large offset P-wave data and model. Lead Edge, 193–198

Hermanrud C, Wensaas L, Teige GMG, Vik E, Nord-gårdBolås HM, Hansen S (1998) Shale porosities from well logs on Haltenbanken (offshore mid-Norway) show no influence of overpressuring. In: Law BE, Ulmishek GF, Slavin VI (eds) Abnormal pressures in hydrocarbon environments, vol 70. AAPG Memoir, pp 65–85

Hoesni MJ (2004) Origins of overpressure in the Malay basin and its influence on Petroleum systems. Ph.D. thesis, University of Durham. Figure 3.10, page 101

John A, Kumar A, Karthikeyan G, Gupta P (2014) An integrated pore pressure model and its application to hydrocarbon exploration: a case study from the Mahanadi Basin, east coast of India. In: Paper appears in Interpretation, vol 2, Society of Exploration Geophysicists and American Association of Petroleum Geologists, pp SB17–SB26

John A, Soni M, Gaur M, Kothari V, AAPG GTW Oil and Gas Resources of India: Exploration and Production Opportunities and Challenges, Mumbai, India, 6–7 Dec 2017

Katahara K (2006) Overpressure and shale properties: stress unloading or smectite-illite transformation? In: Expanded Abstracts, 76th SEG Annual Meeting, New Orleans, 1–6 October, 1520–1524

Lahann R (2002) Impact of smectite diagenesis on compaction modeling and compaction equilibrium. In: Huffman AR, Bowers GL (eds) Pressure regimes in sedimentary basins and their prediction: AAPG Memoir, vol 76, pp 61–72

Lahann RW, Swarbrick RE (2011) Overpressure generation by load transfer following shale framework weakening due to smectitediagenesis. Geofluids 11:362–375

Nwozor KK, OmuduML Ozumba BL, Egbuachor CJ, Onwuemesi AG, Anike OL (2013) Quantitative evidence of secondary mechanisms of overpressure generation: insights from parts of Niger Delta, Nigeria. Pet Technol Dev J 3:64

O'Conner S, Swarbrick RE, Lahann RW (2011) Geologically driven pore fluid pressure models and their implications for petroleum exploration. Introd. Themat. Set: Geofluids 11:343–348

Ramdhan AM, Goulty NR, Hutasoit LM (2011) The challenge of pore pressure prediction in Indonesia's warm neogene basins. In: Proceedings of Indonesian

Petroleum Association, 35th Annual Convention, IPA11-G-141

Swarbrick RE, Osborne MJ, Yardley GS (2002) Comparison of overpressure magnitude resulting from the main generating mechanisms. In: Huffman AR, Bowers GL (eds) Pressure regimes in sedimentary basins and their prediction, vol 76. AAPG Memoir, pp 1–12

Tingay M, Hillis R, Morley C, Swarbrick R, Okpere E (2003) Variation in vertical stress in the Baram Basin, Brunei: Tectonic and geomechanical implications. Mar Pet Geol 20:1201–1212

Tingay M, Hillis RR, Morley C, Swarbrick R, Drake S (2005) Present day stress orientation in Brunei: a snapshot of "prograding tectonics" in a tertiary delta. J Geol Soc 162:39–49

Tingay M, Hillis R, Swarbrick RE, Morley CK, Damit AR (2007) Vertically transferred overpressures in Brunei: evidence for a new mechanism for the formation of high magnitude overpressures. Geology 35:1023–1026

Tingay M, Hillis RR, Swarbrick RE, Morley CK, Damit AR (2009) Origin of overpressure and pore pressure prediction in the Baram Delta Province, Brunei. AAPG Bull 93:51–74. https://doi.org/10.1306/08080808016

Tingay M, Hillis R, Swarbrick R, Morley C, Damit R (2011) Origin of overpressure and pore pressure prediction in the Baram Delta Province, Brunei. Search and Discovery Article #40709

Tingay M, Morley C, Laird A, Limpornpipat O, Krisadasima K, Suwit P, Macintyre H (2013) Evidence for overpressure generation by kerogen-to-gas maturation in the northern Malay basin. AAPG Bull 97:639–672

Toksoz MN, Cheng CH, Timur A (1976) Velocities of seismic waves in porous rocks. Geophysics 41:621–645

Yardley GS, Swarbrick RE (2000) Lateral transfer: a source of additional overpressure? Mar Pet Geol 17:523–537

Global Overpressure Scenario

5

Abstract

Dickinson made a serious attempt to understand the geoscience of pore pressure from his studies in the Gulf of Louisiana. Compaction disequilibrium has been considered classically as the mechanism of overpressure. Subsequently, thermal effects, clay and organic matter transformations and osmosis are the other mechanisms put forward by the authors for overpressure. This chapter reviews worldwide overpressure scenarios—from different continents, countries, (hydro-carbon bearing) rocks of different ages, structures, depths and tectonic and sedimentary regimes.

5.1 Introduction

The geoscience of pore pressure became evident mainly after George Dickinson's work in the Gulf of Louisiana (Dickinson 1951, 1953). While studying large number of wells for hydrocarbon exploration, he found that the top of the abnormal pressure zones are encountered in massive mudrocks lying below a sandy system. Dickinson (1953) commented that compaction of argillaceous sediments (Mukherjee and Kumar 2018) reduces permeability below the pore water expulsion limit. The abnormal pressures in the U. S. Gulf Coast were encountered between 2300 and 2750 m depth (Chapman 1994a, b). With

further advancement of drilling, deeper wells were drilled. Along with this, numerous wells in offshore and onshore discovered more abnormal zones. Workers suggested different constraints such as thermal effects (Barker 1972), clay transformations (Powers 1967; Burst 1969; Bruce 1984) and osmosis. Compaction disequilibrium has been suggested as a dominant mechanism for overpressure generation (Bredehoeft and Hanshaw 1968; Summa et al. 1993). Swarbrick and Osborne (1998) stated that overpressure is not yet a well understood geomechanism. Abnormal pressures were encountered in different parts of the world in different tectonic setup such as the Niger delta of Africa, sedimentary basins around the Borneo in New Guinea, rift basins like the North Sea, etc.

5.2 Global Scenario in Overpressure Zones

Several mechanism and their combinations can produce overpressure. For some younger Tertiary rocks the dominant depositional process is deltaic (Law et al. 1998), where rapid deposition and subsidence of sediments i.e. vertical loading creates overpressure. This is a common phenomenon in the Mississippi, Orinoco and Niger delta regions. Furthermore, in areas like Trinidad (Higgins and Saunders 1967), Papua New Guinea (Hennig et al. 2002), California and Gulf of Alaska, tectonic loading is the common cause for

© Springer Nature Switzerland AG 2020
T. Dasgupta and S. Mukherjee, *Sediment Compaction and Applications in Petroleum Geoscience*,
Advances in Oil and Gas Exploration & Production, https://doi.org/10.1007/978-3-030-13442-6_5

overpressure. Tectonic loading involves many processes like sliding, compression, salt movement and these contribute to overpressure. Wells in the Barbados accretionary prism showed shifting of stress regimes away from the fault (Saito and Goldberg 1997). The U.S. Gulf coast is stated as the type area for abnormal pressures.

In deep waters, the large column of sea water results in deeper overpressure zones, and moderate overpressure zone starts at 9.5 ppg of equivalent mud weight because of the lower fracture gradient. The wells in Gulf of Mexico showed the top of overpressure in thick Pliocene section is ~2118 m below the mud line. In deeper water, the top of overpressure in Miocene section is ~3414 m below mud line in Green Canyon, Garden Banks and Alaminos Canyon. A pressure kick was encountered in the formations below the tabular salt canopy in Walker Ridge and Keathley Canyon. This canopy acts as an effective seal. The lower continental slope consists of a fold and thrust belt, which is prospective and several discoveries have been made (Peel and Matthews 1999; Rowan et al. 2000). The top of the large relief fold and anticline showed that the fluid pressure approached the fracture gradient of adjacent shale (Traugott 1997) and thus showed a centroid effect, where the pressure at the crest in a reservoir sand of a high relief structure exceeds than that of the bounding shale.

The three petroleum provinces of Ukraine: (i) Dnieper-Donets basin, (ii) Carpathian basin and (iii) the North Black sea-provinces are abnormally high pressured. Polutranko (1998) stated that transformation of organic matter and other constituents at >175 °C mainly causes overpressure.

In Pakistan, the abnormally overpressure zones occurs in both Neogene and pre-Neogene sediments. Overpressured Neogene sediments are present onshore and offshore the Makran Basin and the Potwar plateau consist of pre-Neogene sediments (Law et al. 1998). These overpressured zones posed drilling and completion related problems. The Potwar plateau, a part of the Siwalik range (Mukherjee 2015), is the principal

hydrocarbon producing region and the main reservoirs occur in pre-Neogene carbonates and sandstones. The Potwar plateau is bound by faults and N and E, the salt range at S and Indus river in the W. Collision between Indian and Eurasian plates (Mukherjee 2013) produced overpressure in these Neogene rocks. Malick (1979), Sahay and Fertl (1988) and Kadri (1991) mentioned compaction disequilibrium as an alternative cause of high formation pressure. The top of the overpressure zone is at a shallow depth of 290 m in the E Potwar plateau (Law et al. 1998; Malick 1979). The pore pressure in pre-Neogene is less than that in the Neogene sediments and is due to a combination of hydrocarbon generation and tectonic compression.

In the San Andreas fault, overpressure is related to fault geometry and kinematics (Unruh et al. 1992). Fulton et al. (2005) have reported on the presence of a low permeability serpentinite bed across the 50 km wide model domain and there is development of regional overpressure.

Borneo has got few major oil fields such as Sanga-Sanga in Kalimantan and Sarawak. The crude production is from very shallow depth such as <800 m in Bunju, 450 m at Handil, <500 m in Tarakan and the shallowest was at 288 m. The abnormal zones are below the producing horizons (Chapman 1994a, b). Mud volcanoes are associated with these overpressured mudrocks.

In the Niger delta, crude was commercially available since 1955 at Oloibiri (Chapman 1994a, b). Kicks rendered drilling difficult. The Akata Formation occurring at the bottom zones is severely overpressured across the delta and because of the gravitational load of the delta these mud rocks participated in mass flow (Dailly 1976).

Law et al. (1998) compiled ~150 abnormally overpressured locations. 180 abnormally overpressured geographical locations were identified Hunt (1990). Figure 5.1 shows the global distribution of overpressure and underpressure zones.

Bigelow (1994a, b) referred abnormally pressured reservoirs worldwide starting from the Cambrian till the Pleistocene. Different rocks such

Fig. 5.1 Global occurrence of overpressure zones taken from Bigelow (1994a, b) and physical world map of geology.com

as shales, shaly sands and evaporites occurring at both onshore and offshore are abnormally over-pressured (Bigelow 1994a, b). Chapman (1994a, b) stated that the abnormal pressure generation in these reservoirs is not because of compaction disequilibrium, aquathermal expansion or hydro-carbon generation. These pressures are function of densities of the fluid and that of the height above the hydrocarbon contact. Also, worldwide these abnormally pressured reservoirs are closely associated with accumulation of hydrocarbon in both conventional and unconventional reservoirs (Law et al. 1998). These overpressured reservoirs adversely affect the planning as well as financial part of the Petroleum industry.

Pore water overpressure is considered as one of the important reasons for submarine landslides (where slope is <3°). Certain morphological features such as mud volcanoes and pockmarks are associated with submarine landslides.

Table 5.1 Categorization of overpressure zones (Roy et al. 2010)

(A) Overpressure mainly due to compaction mechanism and diagenetic changes	Bengal basin, KG basin, Cambay basin, Bombay Offshore basin
(B) Overpressure generated due to tectonic influence	Himalayan foothills, Jammu area
(C) Overpressure generated due to combination of compaction and later influenced by tectonics and upliftment	Schuppen Belt and adjoining areas of Assam and Arunachal Pradesh, Cachar area of Assam and Tripura-Mizoram fold belts, Andaman Nicobar Basin

5.3 Specific Cases (Table 5.1)

5.3.1 Abnormal Pressure Occurrences in Middle East

The Middle East shows wide variation of structures that contribute to overpressured reservoirs. Overpressure happens in diverse rock/structures such as salt diapirs, massive evaporites and shales. Density contrast amongst fluids also plays an important role in abnormal pressure generation (Bigelow 1994a, b). The formation pressure gradients in the wells of Iraq and Iran ranges from 19.2 to 21.9 kPa m^{-1} and these overpressured wells are drilled with mud weight around 25.8 kN m^{-3} (Bigelow 1994a, b) (Fig. 5.2).

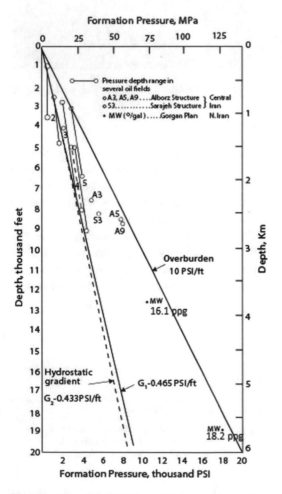

Fig. 5.2 Abnormal formation pressure environments in Iran (Fertl 1976)

Overpressured reservoirs are encountered in several wells drilled in Saudi Arabia, Oman and the United Arab Emirates. High temperature and high pressure wells are also encountered in offshore and onshore wells in the Red Sea area.

5.3.2 Abnormal Pressure Occurrences in Europe

The central graben North Sea (Fig. 5.3) contains one the world's largest Devonian to Early Eocene oil and gas reservoirs. The Central graben of North Sea contain three complex pressure zones viz., (i) highly overpressured Pre-Cretaceous rocks; (ii) Chalk Group with variable pressure regimes; and (iii) normal pressured Paleocene sandstones (Holm 1998).

The hydrocarbons in the supergiant Ekofisk field is produced from overpressured reservoirs of Upper Cretaceous Chalk whereas the Forties field (Fig. 5.4) produces from normally pressured Forties sandstone. Similarly in the Emblafield gas and oil is produced from overpressured pre Jurassic sandstones (Holm 1998). The overpressure pattern in the Viking graben resembles that of the Central graben. In the deepest part overpressured Pre-Cretaceous sediments exist. Compaction disequilibrium is one of the main reasons for overpressure generation in this area. Cayley (1987) studied the overpressured zones at Jurassic top and Paleocene bottom and concluded that the vertical and lateral seals resulted in pressure retention even at basin margins. Cayley (1987) considered undercompaction and aquathermal expansion are the main mechanisms of overpressure. Buhrig (1989) was probably the first to discuss hydrocarbon generation as one of the reasons for overpressure in the North Sea region. Overpressured zones are characterized by gas chimneys produced possibly by vertical seal failure (Buhrig 1989). The Norwegian central graben was divided into three regions (Leonard 1993) viz., the Tertiary, the Chalk group and the pre-Cretaceous. The topmost zone is normally pressured and the overpressure increases towards the lowermost part. The lowermost part is

Fig. 5.3 Location map of the central graben and the quadrant is the licensing structure in the United Kingdom Continental of the area (Holm 1998)

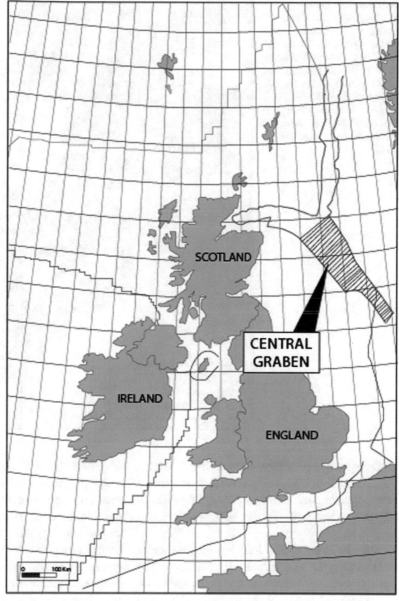

overpressured by 38–52 MPa, and oil to gas conversion created overpressure in the lowermost compartment (Leonard 1993). Gaarenstroom et al. (1993) concluded that rapid sedimentation, fast burial of impermeable claystones and hydrocarbon generation produced overpressure in the central graben area. Generally oil-based mud weight ranging 13–15.5 ppg were used to drill wells in Norway and United Kingdom to control the high pressured zones as well as gas

pockets (Bigelow 1994a) (Figs. 5.5 and 5.6). Along with these mechanisms of overpressure generation, Holm (1996) addressed the failure of the seal in the overpressured zones.

The methods of overpressure prediction in the North Sea resemble that of United States in terms of integrating seismic data, drilling and logging parameters used for the analysis and pre-drill overpressure prediction matches the post-drill one (Bigelow 1994a) (Fig. 5.7).

Fig. 5.4 The Forties
Montrose High divides the
graben into eastern and
western grabens. The oil and
gas fields are from Devonian
to early Eocene age and the
major formations are Upper
Jurassic Fulmar Sandstone,
Upper Cretaceous Chalk
Groupand the Paleocene
Forties Formation (Holm
1998)

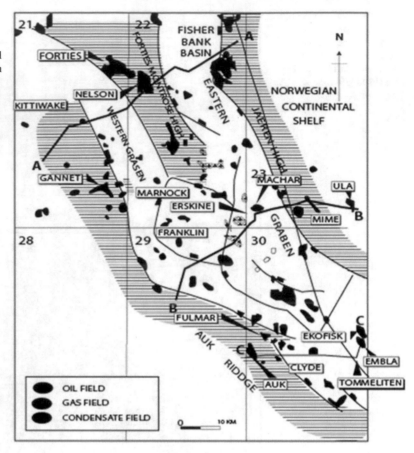

Overpressured shales and highly complex folded, faulted hydrocarbon bearing sequences were encountered during drilling of wells in offshore Ireland Celtic sea (Bigelow 1994a) (Fig. 5.8).

Two tectonic events lead to complex structural setting in the Italian Adriatic area. The Apulian passive continental margin is the result of extensional tectonics (Mattavelli et al. 1991) and compressional tectonics gave rise to the Apennine Belt. Figure 5.9a is the simplified version of the tectonic map showing the location of the northern and central Adriatic Basin, and the major structural features of the Apennines (Italy). Five overpressured zones are identified here: three in the post-Messinian siliciclastic succession of the Adriatic foredeep and two in the carbonates of Miocene to Cretaceous age of the Apulian continental margin. Drilling-related problems, blowouts, casing collapse etc. were

frequently encountered in onshore and offshore wells (Bigelow 1994a). Figure 5.9b shows the Pressure-depth plot illustrating the terminology used to describe pressure regimes. The figure shows a hypothetical pressures profile of a well (defined by points R through Y). The three main types of pressure regimes that can be identified in strata of the northern and central Adriatic Basin are shown in this figure. A normal-pressure regime is indicated by A, an overpressured regime with a hydrostatic gradient is indicated by B, and an overpressured regime with overpressure gradient is indicated by C. Some of overpressure-related problems were avoided using pressure evaluating techniques and shows offshore wells in the Adriatic sea. Figure 5.9c is the combination of different parameters including drilling, logging etc., which indicated the presence of overpressure environment in the Adriatic sea. The thrusts present at the boundaries of these

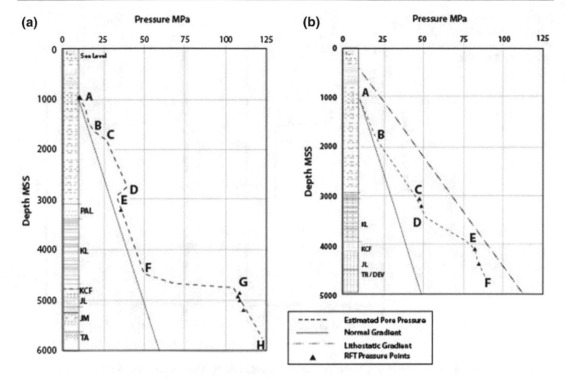

Fig. 5.5 Pressure versus depth plots. The measured pressure data from the permeable zones have been used and in the impermeable zones the pressures have been estimated by indirect methods of pressure estimation (mud-weights, D-exponents, connection gases, etc.). **a** Pressure versus depth plot in the Central graben area which shows the normally pressured Paleocene sandstones and beneath that there is normally pressured chalk group. There is rapid change in the profile with increase in the pore pressure from **f** to **g** and the highest at the pre-Cretaceous structure in the **h**. **b** Pressure profile of Ekofisk Field. The chalk group (**c**, **d**) is moderately overpressured (Holm 1998)

regions act as pressure barriers. Compaction disequilibrium is the main reason for the presence of abnormally overpressured zones (Carlin and Dainelli 1998).

5.3.3 Abnormal Pressure Occurrences in Africa

In the African continent, abnormal pressures are encountered in wells from several countries. Overpressured zones are present in both the E and W coasts of Africa but the overpressured zones are more pronounced along the E coast (Bigelow 1994a). Wells in Mozambique, Madagascar and the offshore Red Sea show abnormal pressure zones.

Eight pressure compartments with lateral and vertical seals were identified in the Nile delta and in the North Sinai Basin (Nashaat 1998). Abnormally high pressure was defined in the Nile Delta and North Sinai Basin up to $\sim 16,000$ ft (4900 m) depth (Fig. 5.10).

The top of overpressured zones vary from 520 to 3700 m and the pressure gradient is as high as 20.2 kPa m^{-1}, which differs from the normal pressured reservoir gradient of 9.95 kPa m^{-1}. Areas with high sedimentation showed abnormal pressures even at shallow depths in the Pliocene reservoirs and the sedimentation rate in these areas is as much as 70 cm per 1000 year (Nashaat 1998). This high rate of sedimentation can trigger compaction disequilibrium and abnormally high pressures (Mann and Mackenzie 1990).

Secondary mechanisms of overpressure including aquathermal expansion, hydrocarbon

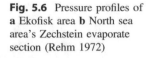
Fig. 5.6 Pressure profiles of
a Ekofisk area **b** North sea
area's Zechstein evaporate
section (Rehm 1972)

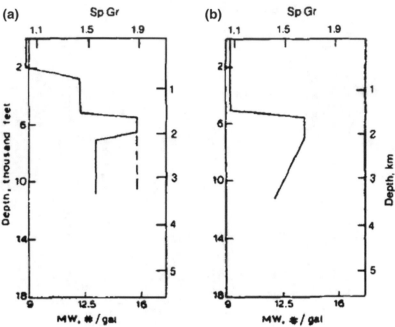

generation and oil to gas cracking were responsible for abnormal pressure in the Upper Oligocene and Lower Miocene sections of the eastern
Nile delta (Nashaat 1998). Evaporite strata of
Messinian age (Uppermost Miocene) develops
pore pressure in the pre-Messinian part by pro-
hibiting pore fluid flow (Nashaat 1998).

5.3.4 Abnormal Pressure Occurrences in China

Geological conditions suits overpressured reservoirs for >200 Chinese basins (Bigelow 1994a).
Complex geological structures such as thrusts,
compressional folds, nappes, en-echeleon faults,
horst and grabens are predominant in several
parts of China and these formed by plate movements (Bigelow 1994a).

The Yinggehai Basin is a high-temperature
and high-pressure, rift to passive margin basin
located at the NW South China Sea (Fig. 5.11)
(Hao et al. 1998, 2000). The common overpressure generating mechanisms are disequilibrium
compaction, smectite to illite transformation,

aquathermal pressuring and kerogen maturation
(Liu 1993; Zhang et al. 1996; Hao et al. 2000).

The top of overpressured zone is as deep as
4000 m towards the basin margin and overpressure increases towards the basin center where it
occurs between 1500 and 2400 m depth. Luo
et al. (2003) referred to an allogenic mechanism
of pressure generation where the overpressured
zones occur at shallow depths and the source of
abnormal pressure is deep-seated. When differential pressure persists in two different reservoirs,
which are separated by open faults, there is quick
pressure readjustment and this created an overpressured system in the Yinggehai Basin
(Luo et al. 2003).

One of the most hydrocarbon prolific basin is
the Bohai Bay basin with many sub-basins, which
is located in the E coast of China (Guo et al.
2010). The Dongying sub-basin part of Bohai
Bay basin is filled with Cenozoic sediments
consisting of mixed lithologies of mudstones,
medium-grained sandstones and siltstones
(Guo et al. 2010).

The pressure coefficient method was applied
to identify the abnormal zones in Dongying

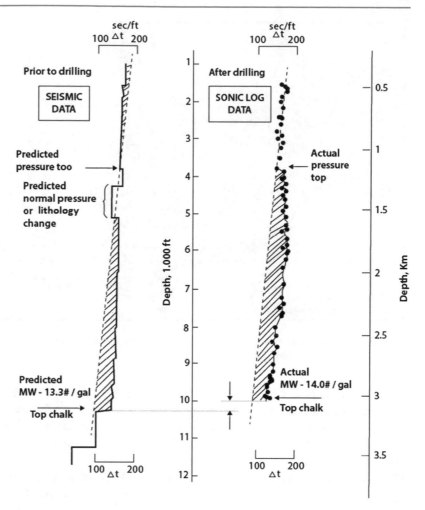

Fig. 5.7 Comparison of predrill and actual pore pressure prediction in north sea area (Herring 1973)

sub-basin. The pressure coefficient is the ratio of actual pore pressure recorded to that of normal hydrostatic pressure (Guo et al. 2010). Pressure in different stratigraphic units of different formations were obtained from 1400 sandstone data and were plotted against depths to delineate the abnormal pressure zones (Fig. 5.12). Es3 and Es4 Formation are overpressured as indicated by the cross-plot between pressure coefficient and depth where the top of overpressure occurs between 2000 and 3800 m in both formations. The pressure coefficients range between 0.9 and 1.95 with maximum in Es3 and Es4 Formation. Rest of the formation seemed to be normally pressured from the plot.

Sonic, density, resistivity wireline logs were studied in Dongying depression. Guo et al.

(2010) suggested the sonic logs can reliably point out overpressured zones in the Dongying depression because the interval transit time increases within the overpressured mudstones. Interpretations were supported by pressure coefficient results and mud weights. Distinct changes in the interval transit time were recorded in the well Bin670 (Fig. 5.13) where the top of overpressured zone is encountered at 2550 m depth. Resistivity values reduce in overpressured sediments (Hermanrud et al. 1998). But the typical resistivity log signatures of overpressured sediments were not found in case of the Dongying depression. The zones in the well delineated as overpressured zones with higher interval transit time did not show reduced resistivity values when compared to the normally pressured ones.

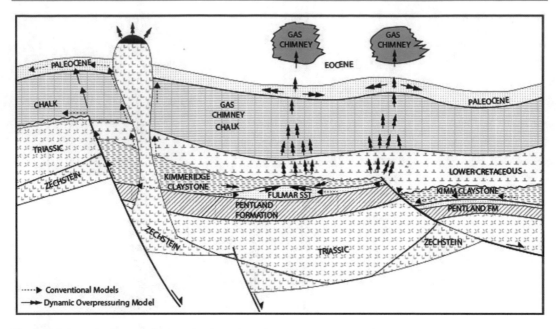

Fig. 5.8 Schematic diagram depicts the formation of overpressure and underpressure zones as the hydrocarbon migration takes place (Holm 1998)

As stated in Chap. 4, the response of resistivity log depends upon factors such as minerals, type of fluid present in the formation and also the fluid retention. Zones with high organic content in the overpressured part in the Dongying depression displayed higher resistivity values. Similarly lowering of density against the overpressured zones was not observed in the Dongying basin (Guo et al. 2010).

In the Sichuan basin of China, pressure data from Permian carbonate reservoirs yielded a pressure gradients of >14.2 kPa m^{-1}.

The northern part of South China sea has the highest hydrocarbon potential and deep formations like Enping and Wenchang Formations are overpressured in Baiyun sag (Kong et al 2018). In the latter sag basin the rapid sedimentation is considered to be the main cause of overpressure generation apart from hydrocarbon generation and fault activity is considered to be the cause of pressure release (Kong et al. 2018).

5.3.5 Abnormal Pressure Occurrences in and Around SE Asia, Australia and New Zealand

The W part of Taiwan is extremely folded as it belongs to foreland fold and thrust belt (Suppe and Wittke 1977) and Hubbert and Rubey (1959) stated that fluid plays an important role in the mechanical properties of porous rocks.

Thrusting may occur if the pore fluid pressure nearly equals normal stress. The Hubbert and Rubey's theory explains movement of the strike slip fault at Rangely, Colorado. Fluid pressure data were collected by the Chinese Petroleum Corporation in the complex structures of NW part of Taiwan (Fig. 5.14) show the relationship of the fluid pressure versus depth ratio with stratigraphic horizon. Chuhuangkeng and the Pakuali areas exhibit abnormal pressure at ~2.8–4.1 km depths, which follow the hydrostatic pressure

Fig. 5.9 **a** Tectonic map of the Italian Adriatic area. **b** Pressure profile of a well in the Italian Adriatic area. In the zones B & C the pressure values are far beyond the normal pressure values. **c** Plots showing overpressured environments in Adriatic sea (Rizzi 1973)

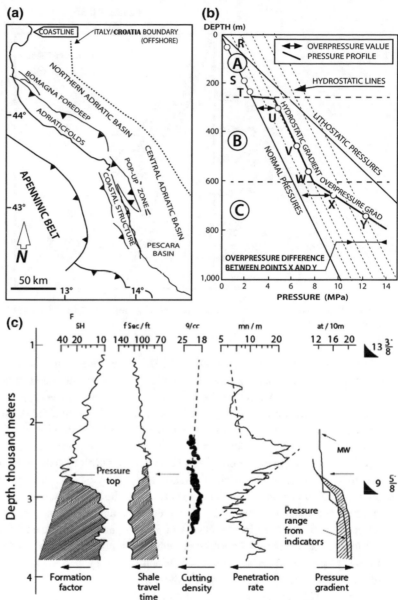

gradient. At Tiechenshan and Chingtsaohu-Chiting, abnormal fluid pressures occur at >5 km depth.

Different parts of the Australian basins show variation in pore pressure (Fig. 5.15). For example the Perth Basin, Bonaparte Gulf Basin and the Carnarvon Basin, in both onshore and offshore areas show variation of pore pressure (Bigelow 1994a).

The pressure gradient in shallow wells of Queensland is ~ 13.6 kPa m^{-1} and the magnitude is

~ 12.44 kPa m^{-1} on the Gippsland shelf. Blowouts are also reported from wells from these areas (Bigelow 1994a). The mud weights commonly used in these overpressured formations in W as well as N Territory of Australia exceeds 16.5 kN m^{-3}. An abnormal pressure gradient of 0.65 psi/ft is reported from the island of Timor and these zones occur at shallow depths (Bigelow 1994a).

Tertiary and Mesozoic sediments of the Papua New Guinea are overpressured. The pressure

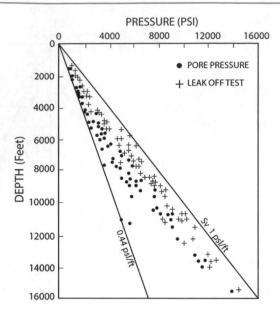

Fig. 5.10 Pore pressure plot of the North Sinai basin and the Nile delta. Abnormal increase in the pore pressure and leak off test is there (Nashaat 1998)

gradients in the Aure Trough is ~ 20.6 kPa m^{-1} and the hydrocarbon bearing limestone reefs in these areas have high pressure ~ 14 kPa m^{-1}.

Hydrocarbon exploration in parts of Indonesia started during the early 19th century. Some fields are producing from shallow depths. For example, the Seria field in Brunei is producing at 288 m, $\sim$450 m at Handil, <500 m at Tarakan, and <800 m at Kalimantan (Chapman 1994a, b). Abnormally pressured mudrocks are present beneath productive zones in the basins of Borneo. Mud volcanoes are also associated with the abnormally pressured zones such as around southern Sabah, Setap shale beneath the producing zones (Liechti et al. 1960). Abnormally pressured fluids exist in the reservoirs in the Baram delta (Sarawak; Schaar 1977). After Sumatra, lower Kutai is the second largest hydrocarbon producing field in Indonesia (Ramdhan et al. 2011). Pliocene marine muds and mudstones dominate the offshore region of the lower Kutai Basin and the shelfal region supplied sediments to the deep water turbidite reservoirs. The onshore area is dominated by sandy facies, which is uplifted and eroded at places. Mixed lithological sequences consisting

of sandy and muddy facies are present in the shelfal area and this sequence is fed by the Mahakam river.

Rapid sedimentation and quick burial created overpressure by compaction disequilibrium in the Neogene sediments (Burrus 1998). Ramdhan and Goulty (2010) concluded from the data of a single field that the overpressure generation in the shelfal area is because of secondary processes such as unloading, gas generation and clay diagenesis. Subsequent studies revealed that the entire basin is overpressured (Ramdhan et al. 2011).

5.3.6 Abnormal Pressure Occurrences in Some Parts of Asia

Overpressured zones are encountered in many parts of India, Bangladesh, Pakistan and Myanmar (Bigelow 1994a). In India the overpressured zones are present in offshore Krishna-Godavari Basin, Bengal Basin, Schupen belt of Assam, Kutch, Mizoram and Tripura, Himalayan foothills, offshore areas of Bombay/Mumbai and Andaman Basin (Fig. 5.16). Sahay (1999)

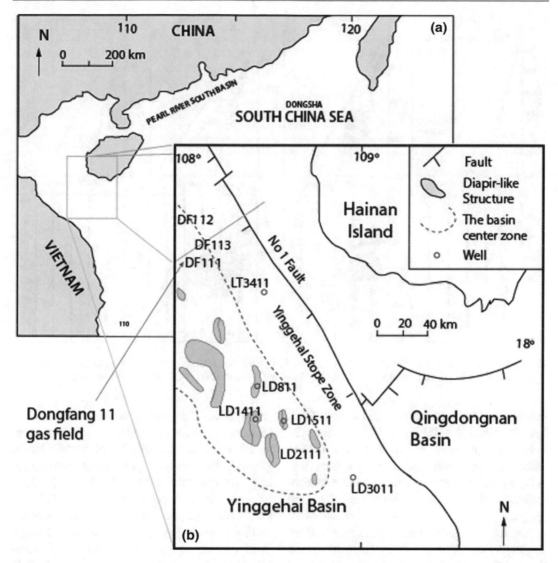

Fig. 5.11 Location of Yinggehai basin in the South China sea along with the location of wells along with structural features such as faults, diapir like structures

referred briefly the overpressured zones in India sedimentary basins. We are not aware of over-pressure cases from Indian inland basins.

In his review Roy et al. (2010) stated that the exploration in Bengal Basin started in 1957–1960 by Indo Stanvac Petroleum Company (ISPP). Out of the ten wells drilled, one well (W7) was drilled up to 4042 m depth and over-pressure was encountered below 4030 m as indicated by usage of high mud weight with specific gravity 1.71 g/cc. In the deepest basinal

part, ONGC drilled its first well W-10. Over-pressure was encountered at 3775 m depth in this well while drilling Miocene sand. Later in 1975, another well was drilled to explore fault closure and encountered overpressure at 3540 m depth. Later several wells W22, W26 and W29 were drilled by ONGC and encountered overpressure in Miocene sediments. Growth faults were drilled by ONGC through the wells W30 and W32, which also encountered overpressure. Later in 2005, well W45 was drilled in the slope fan

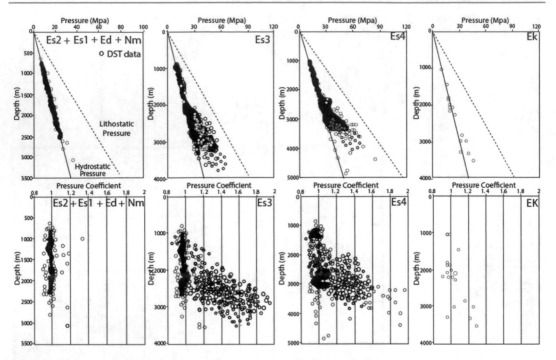

Fig. 5.12 Pressure and Pressure coefficient versus depth plot of the different stratigraphic units in the Dongying depression of the Bohai Bay basin where the Es3 and Es4 are overpressured. DST denotes drill stem testing

prospect and the well was targeted up to 5000 m but terminated at 4400 m due to hole complications as overpressure was encountered at 4278 m where the equivalent mud weight used was ~2.12 g/cc sp. gravity and the well was controlled with a mud weight of 2.28 g/cc sp. gravity.

The log signatures of an overpressure well are shown in Figs. 5.17 and 5.18. The sonic transit time (ΔT) is plotted with depth and normal compaction trend is built for each of the well. The departure from the normal trend is taken as the transition zone and is confirmed with the reservoir formation tester data. The cross-plot of shale acoustic parameter difference versus reservoir fluid pressure gradient was made by Hottman and Johnson (1965) for the Oligocene–Miocene formations of the U.S. Gulf coast area. Incorporating further well data, Roy et al. (2010) later modified the Hottmans curve (Fig. 5.19).

John et al. (2014) made an integrated study of pore pressure model and seismic velocities to interpret the pressure distribution along the stratigraphy and its dependence for hydrocarbon

exploration. The study was undertaken for the Pliocene and Miocene reservoirs of the northern to southern sectors of the Mahanadi Basin. Towards the northern part, wells encountered high overpressure >19.7 MPa km^{-1} in Late Miocene strata. A pore pressure of <13.7 MPa km^{-1} was encountered in the Late Miocene section towards the southern part of the basin. Details about the overpressure generating mechanism and its detection from wireline logs in Mahanadi Basin were discussed in Chap. 4. John et al. (2014) tried to correlate the vertical effective stress distribution and its relation to hydrocarbon accumulation. Vertical effective stress from 3D pore pressure model shows a correlation with the reservoir fluid type. It has been observed from the well data that below a cut-off of 7.9 MPa km^{-1} the dominant fluid is water (Fig. 5.20). Except hydrocarbons existing in thin calcareous beds having lower effective stress, tight water bearing reservoirs have higher effective stress. The benefit of this type of study integrating geological and geophysical attributes increases the probability of getting thicker

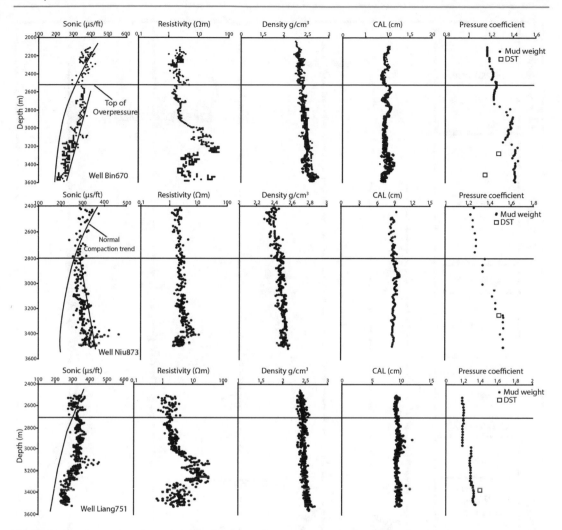

Fig. 5.13 Depth versus sonic, resistivity, density, calliper profile of well log data of wells in Dongying depression

hydrocarbon bearing reservoirs in Miocene-Pliocene reservoirs and that aids in selecting new drilling locations.

Krishna-Godavari basin (India): Another proven petroliferous basin from the E coast of India is the Krishna-Godavari Basin. Hydrocarbon occurs here from Permian to Pliocene reservoirs (Rao 2001). Fault-controlled ridges, ~5 km thick, divide this basin into different sub-basins. Jurassic sediments deposited in the rift valley and were later overlain by transgressive deposits, which was later followed by sedimentation by delta progradation (Rao 2001). Around 8 km thick pile of sediments is

demarcated by geoscientific surveys. Hydrocarbons are distributed in different types of plays including Tertiary deltas as well as deep water channel-levee plays (Bastia et al. 2006).

Singha and Chatterjee (2014) studied the reservoir formation tester data (RFT) from 10 wells (Fig. 5.21) from the eastern continental margin in the Krishna-Godavari Basin, which revealed that the Vadupuru shale (Miocene) and the Raghavpuram shale (Early Cretaceous) are overpressured. The Pre-drill pore pressure model matches the actual reservoir formation tester (RFT) data. The wells are distributed at the MDP, END, RAN, KAV gas fields towards the

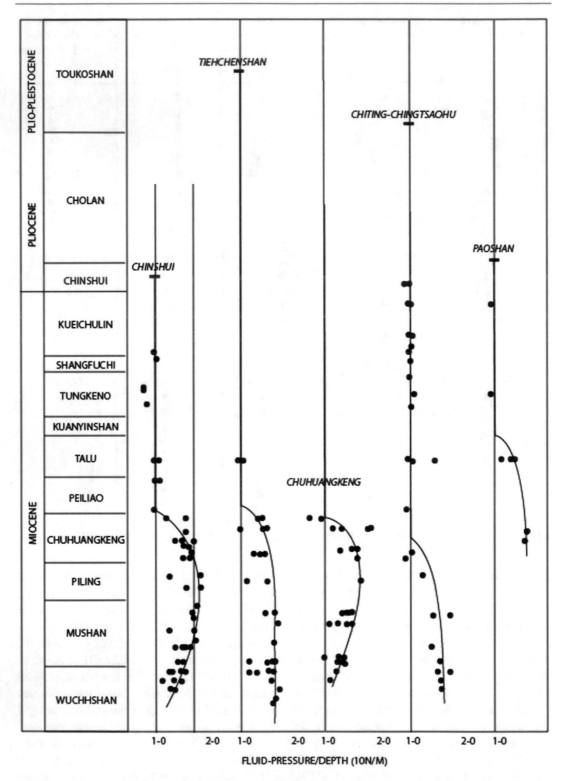

Fig. 5.14 Shows the relationship of the fluid pressure versus depth ratio with stratigraphic horizon. The stratigraphic zones in different wells are overpressured because of the presence of apparent low permeability zone

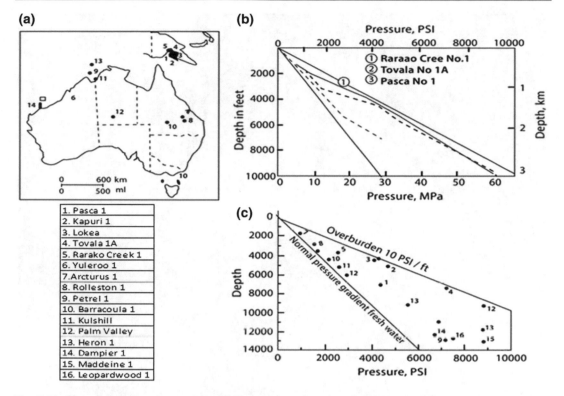

(a)

1. Pasca 1
2. Kapuri 1
3. Lokea
4. Tovala 1A
5. Rarako Creek 1
6. Yuleroo 1
7.Arcturus 1
8. Rolleston 1
9. Petrel 1
10. Barracoula 1
11. Kulshill
12. Palm Valley
13. Heron 1
14. Dampier 1
15. Maddeine 1
16. Leopardwood 1

Fig. 5.15 Figures showing abnormal formation pressures in Australia and Papua New Guinea area. **a** Location of overpressured wells. **b** Pressure profile of overpressured wells of Australia. **c** Pressure profile of overpressured wells of Papua New Guinea. (Bigelow 1994a, b)

eastern part of the Godavari sub-basin and MDH, SUR fields towards the western part of the Godavari sub-basin (Singha and Chatterjee 2014; Chatterjee et al. 2015) (Fig. 5.22).

For the delineation of the overpressured zones, sonic logs from these wells were used for pore pressure predication in shales of Krishna-Godavari Basin from popular methods such as Eaton and equivalent depth method from sonic velocity (Van Ruth et al. 2002). The main objective of Singha and Chatterjee (2014) and Chatterjee et al. (2015) was to delineate the overpressure zones by establishing the normal compaction trend (NCT) from sonic logs and the deviation of the data from the NCT for the wells of Krishna-Godavari Basin. Chatterjee et al. (2011) identified that the overpressured zones are confined to low porosity medium and NCT was established in shales (Singha and Chatterjee 2014; Chatterjee et al. 2015). The sonic transit time

(DT) as well as porosity from density as well as sonic log were plotted for two wells (#7 and #13) (Fig. 5.23). The top of the overpressured zone is indicated by sonic data deviating from NCT as well as from the separation between the sonic porosity and density porosities.

Kutch Basin (India): According to Sahay (1999), overpressure was encountered in drilling an exploratory well during 1975 in the Kutch Basin. High pressure ranging of 389–545 kg cm^{-2} was encountered at 2600 m up to the target depth at 4575 m. The pore fluids were mainly water and gas and the overpressure rise ranged 1.2–1.4 times that of the hydrostatic pressure. The variation in pressure, temperature and salinity are presented in Fig. 5.24.

Western basins of India-Saurashtra Dahanu block (Western offshore Basin) and Cambay Basin: In the Bombay/Mumbai offshore area and N to the Bombay platform, the Saurashtra,

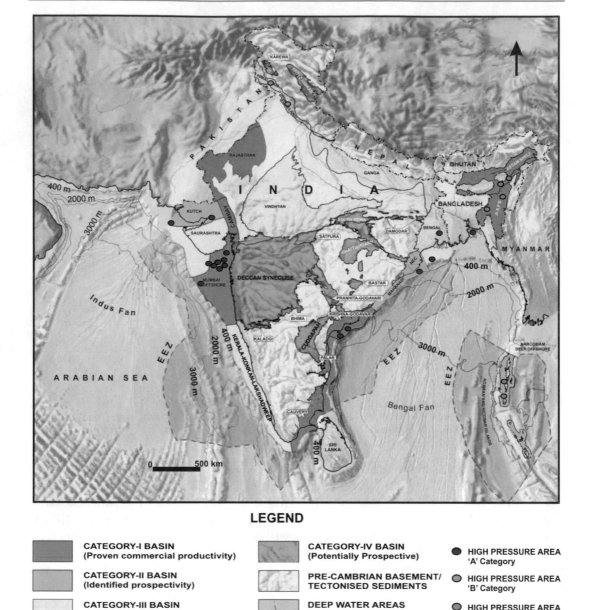

LEGEND

![CATEGORY-I BASIN]	**CATEGORY-I BASIN** (Proven commercial productivity)	![CATEGORY-IV BASIN]	**CATEGORY-IV BASIN** (Potentially Prospective)
![CATEGORY-II BASIN]	**CATEGORY-II BASIN** (Identified prospectivity)	![PRE-CAMBRIAN]	**PRE-CAMBRIAN BASEMENT/ TECTONISED SEDIMENTS**
![CATEGORY-III BASIN]	**CATEGORY-III BASIN** (Prospective Basins)	![DEEP WATER]	**DEEP WATER AREAS WITHIN EEZ**

● **HIGH PRESSURE AREA** 'A' Category

◐ **HIGH PRESSURE AREA** 'B' Category

◑ **HIGH PRESSURE AREA** 'C' Category

Fig. 5.16 Sedimentary basins of India showing overpressure area (Sahay 1999)

Dahanu PEL block is situated covering ~ 2500 km^2 (Fig. 5.25).

Overpressure in this area occurs within 1800–3600 m depth in Early Eocene to Early Miocene shales and carbonates (Nambiar et al. 2011). Seismic interval velocity predicted the overpressure before drilling wells here. Because of limited formation pressure data, Nambiar et al.

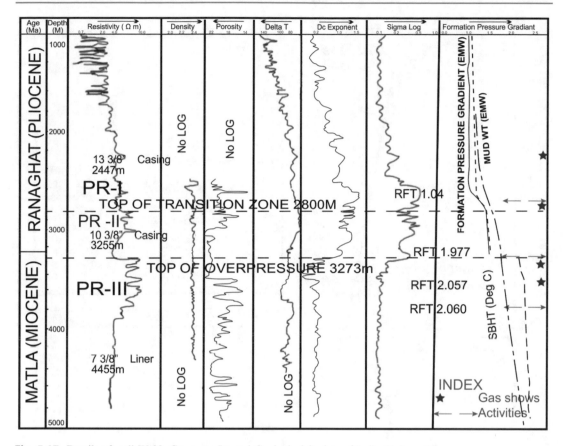

Fig. 5.17 Details of well W-32. Courtesy: Journal Geological Society of India, Volume 75

(2011) used mud weight data as an indicator for overpressure. In these high pressure-high temperature wells, high mud weight was used to control the well events.

The interval velocities of the wells B-9-A, B-9-B, B-9-C, B-9-D, B-12-9-A and B-12-9-B are shown in Fig. 5.26. Except for well B-9-D, all the wells showed lowering of interval velocity and no overpressure was encountered while drilling well B-9-D (Nambiar et al. 2011). In well B-12-9-A, the formation pressure encountered is 16.4 ppg mud weight equivalent and 16.8 ppg mud was used to control the well (Nambiar et al. 2011).

Pakistan: Abnormal fluid pressure is encountered in the Himalayan foothills at Potwar, Pakistan. Overpressure complicated drilling and completions of oil and gas wells (Law et al. 1998). Besides, the Makran Basin in the southern Pakistan is also overpressured (Fig. 5.27).

High pressures have been encountered in drilling Neogene rocks in the Potwar plateau area (Sahay and Fertl 1988). The main causes of overpressure generation were (i) India-Eurasia collision (Sahay and Fertl 1988); and (ii) compaction disequilibrium (Kadri 1991). Law et al. (1998) suggested the presence of abnormal pressures also in Paleogene and older rocks.

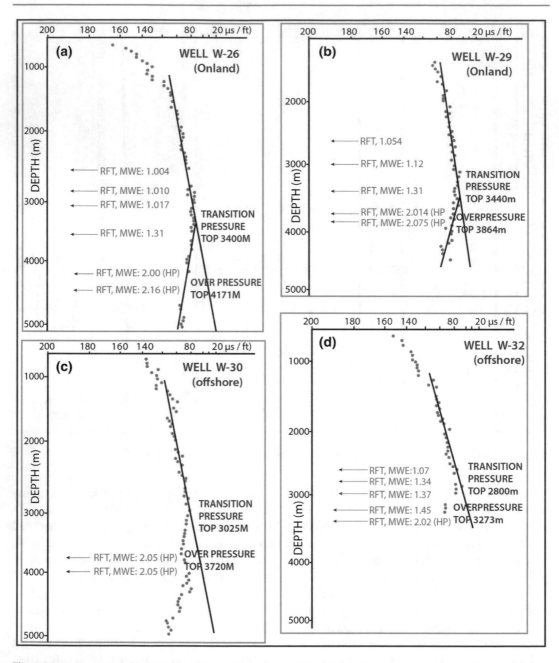

Fig. 5.18 Depth versus ΔT plot of the wells in Bengal basin, the deviation from the normal compaction trend is taken as the top of the overpressure and is also confirmed with the RFT results

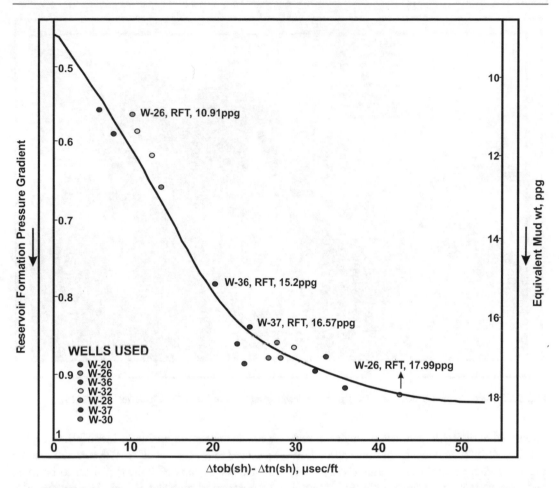

Fig. 5.19 Relationship between shale acoustic parameter difference $\Delta tob(sh) - \Delta tn(sh)$ from drilled wells and reservoir fluid pressure gradient (modified Hottmans curve of Bengal Basin) (Roy et al. 2010)

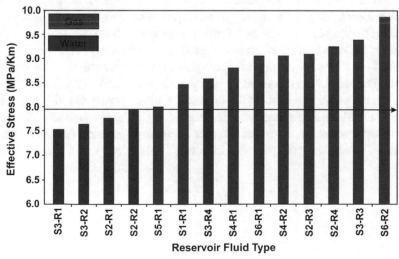

Fig. 5.20 Relation between effective stress from well data and reservoir fluid type in which below 7.9 MPa km^{-1} cut off the chances of getting hydrocarbon is minimum (John et al. 2014)

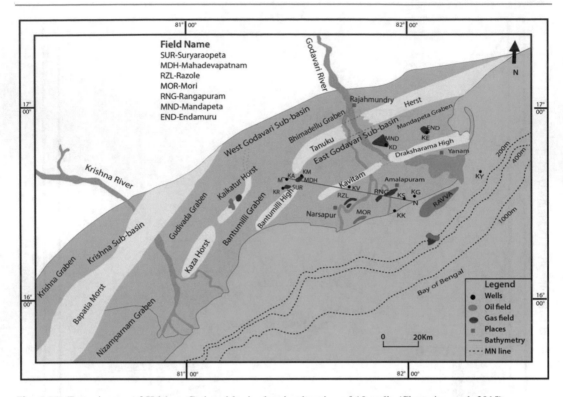

Fig. 5.21 Tectonic map of Krishna–Godavari basin showing location of 10 wells (Chatterjee et al. 2015)

Along with different parameters such as shale density, drilling exponent, drilling speed, flow-line temperatures, drill stem testing (DST) results and mud weights were commonly considered as overpressure indicators (Law et al. 1998) (Fig. 5.28).

In certain areas sonic transit time were also considered as a tool to detect pressure (Fig. 5.29). Mud weight and drilling exponent data indicated average pressure gradients of 20.4–22.6 kPa m^{-1}. The actual pore pressures in the Potwar plateau could not be estimated based on mud weights as most of the operators drilled wells with overbalanced mud because of high pressure scenarios, so minor changes in hydrostatic changes could not be detected. Thus,

geoscientists were unable to identify lithological pressure seals. Law et al. (1998) reported that the N and the E parts of Potwar plateau to be more overpressured than the S part. The authors also found a relationship between the geological structures and overpressure development.

Law et al. (1998) studied the Neogene and Pre-Neogene pressure systems separately. They made few important observations during their study. The drill stem testing (DST) results showed that Pre-Neogene sediments are more saline (7000–41,000 ppm) than Neogene sediments (2500–5700 ppm). The Paleogene rocks are organically matured (1.5–30% total organic content) than that of the Neogene rocks (organically <0.5% organic matter). Based on different

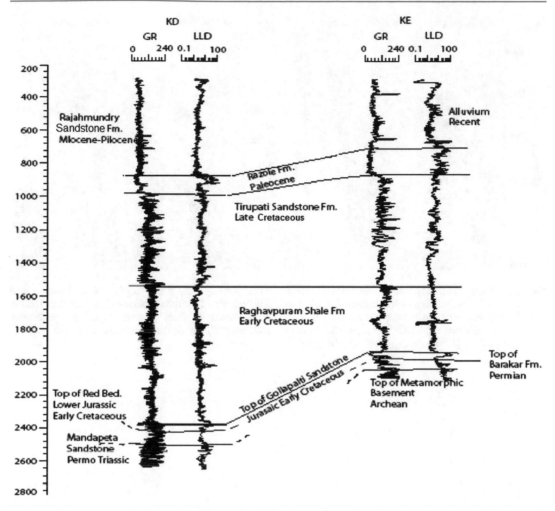

Fig. 5.22 Gamma ray and resistivity log responses of wells KD & KE. Location of wells and profile is shown in Fig. 5.21 (Singha and Chatterjee 2014)

studies such as thermal maturity, formation water quality, temperature, sediment deposition rates, structural features etc., Law et al. (1998) concluded that the abnormal pressures in the Neogene sediments are mainly due to tectonic compression and undercompaction whereas the pore pressure in the pre-Neogene sediments is caused by a combination of hydrocarbon generation and tectonic compression.

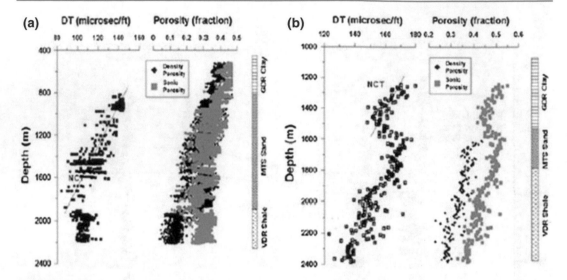

Fig. 5.23 **a** DT versus depth and density porosity sonic porosity (Øs) versus depth plot for onshore well #7. The top of overpressured zone is indicated at 1919.16 m by deviation of sonic data from NCT and separation between (Ød) and (Øs). **b** DT versus depth and density porosity sonic porosity (Øs) versus depth plot for offshore well #13. The top of overpressured zone is indicated at 1600 m by deviation of sonic data from NCT and separation between (Ød) and (Øs) (Chatterjee et al. 2015)

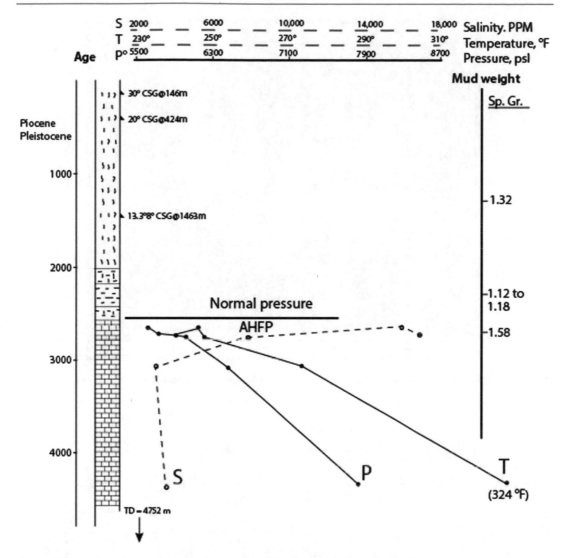

Fig. 5.24 Pressure, salinity and Temperature data in the offshore well Kutch basin, India (Sahay 1999)

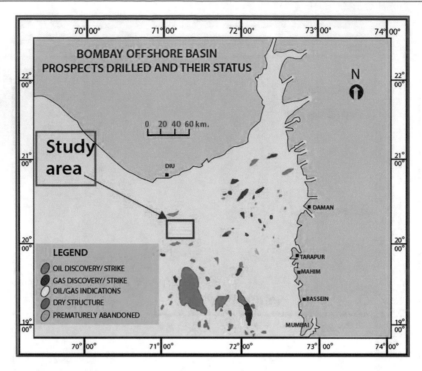

Fig. 5.25 Bombay/Mumbai offshore basin with the location of Saurashtra-Dahanu block (Nambiar et al. 2011)

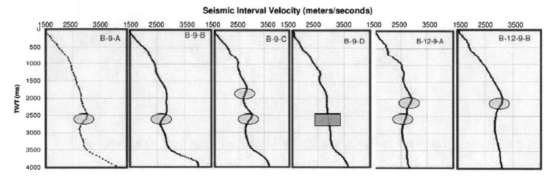

Fig. 5.26 Interval velocities of different wells plotted and the different zones with lowering of interval velocities are marked with circles and denote overpressure zones (after Nambiar et al. 2011)

Fig. 5.27 Location of overpressured zones in Pakistan: Potwar Plateau (**a**) and Makran basin (**b**) Modified after Law et al. (1998) incorporating Google Earth imagery

Fig. 5.28 Pressure and
Temperature gradient in Gulf
oil, FimKhassar well (Law
et al. 1998)

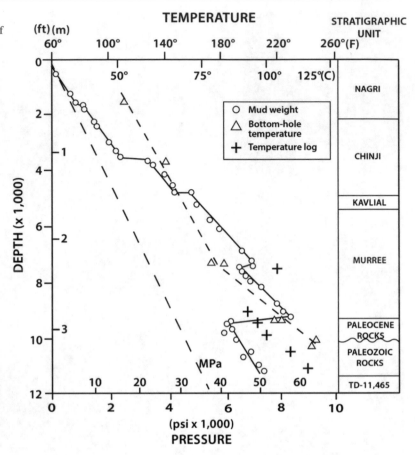

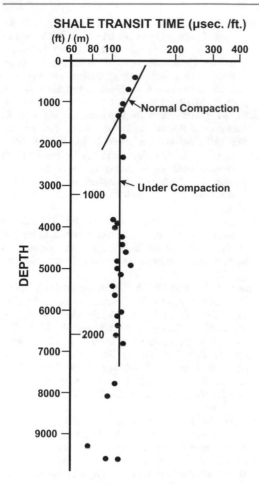

SHALE TRANSIT TIME (μsec. /ft.)

Fig. 5.29 Interval transit time versus depth in the Gulf oil, FimKassar well (Law et al. 1998)

References

Barker C (1972) Aquathermal pressuring-role of temperature in development of abnormal-pressure zones. AAPG Bull 56:2068–2071

Bastia R, Singh P, Nayak PK (2006) Linking shelf delta to deep water; Krishna-Godavari basin. J Geol Soc India 67:619–628

Bigelow EL (1994a) Global occurrences of abnormal pressures. In: Fertl WH, Chapman RE, Hotz RE (1994) Studies in abnormal pressures. Development in petroleum science, vol 38. Elsevier Science, Amsterdam, pp 1–17

Bigelow EL (1994b) Well logging methods to detect abnormal pressure. In: Fertl WH, Chapman RE, Hotz RE (eds) Studies in abnormal pressures. Elsevier, Amsterdam, pp 187–240

Bredehoeft JD, Hanshaw BB (1968) On the maintenance of anomalous fluid pressure, I. Thick sedimentary sequences. Geol Soc Am Bull 79:1097–1106

Bruce CH (1984) Smectite dehydration—its relation to structural development and hydrocarbon accumulation in northern Gulf of Mexico basin. AAPG Bull 68:673–683

Buhrig C (1989) Geopressured Jurassic reservoirs in the Viking Graben: modeling and geological significance. Mar Pet Geol 6:31–48

Burrus J (1998) Overpressure models for clastic rocks, their relation to hydrocarbon expulsion: a critical reevaluation. In Law BE, Ulmishek GF, Slavin VI (eds) Abnormal pressures in hydrocarbon environments: AAPG Memoir 70, pp 35–63

Burst JF (1969) Diagenesis of Gulf Coast clayey sediments and its possible relation to petroleum migration. AAPG Bull 53:73–93

Carlin S, Dainelli J (1998) Pressure regimes and pressure systems in the Adriatic foredeep (Italy). In: Law BE, Ulmishek GF, Slavin VI (eds) Abnormal pressures in hydrocarbon environments: AAPG Memoir 70, pp 145–160

Cayley GT (1987) Hydrocarbon migration in the Central North Sea. In: Brooks J, Glennie K (eds) Petroleum geology of north west Europe. Graham and Trotman, London, pp 549–555

Chapman RE (1994a) Abnormal pore pressures: essential theory, possible causes, and sliding. In: Fertl WH, Chapman RE, Hotz RF (eds) Studies in abnormal pressures. Developments in petroleum science, vol 38. Elsevier, pp 51–91

Chapman RE (1994b) Geology of abnormal pore pressures. In: Fertl WH, Chapman RE, Hotz RF (eds) Studies in abnormal pressures. Developments in petroleum science. Elsevier, pp 19–49

Chatterjee R, Mukhopadhyay M, Paul S (2011) Overpressure zone under the Krishna Godavari offshore basin: geophysical implications for natural hazard in deeper-water drilling. Nat Hazards 57:121–132

Chatterjee R, Paul S, Singha D K Mukhopadhyay M (2015) Overpressure zones in relation to in situ stress for the Krishna-Godavari basin, eastern continental margin of India: implications for hydrocarbon prospectivity. In: Mukherjee S (ed) Petroleum geosciences: Indian contexts. Springer, Springer International Publishing

Dailly GC (1976) A possible mechanism relating progradation, growth faulting, claydiapirism and overthrusting in a regressive sequence of sediments. Bull Can Pet Geol 24:92–116

Dickinson G (1951) Geological aspects of abnormal reservoir pressures in Gulf Coast region of Louisiana, U.S.A. In: Proceedings of the 3rd world petroleum congress, The Hague, Section 1, pp 1–16

Dickinson G (1953) Geological aspects of abnormal reservoir pressures in Gulf Coast Louisiana. AAPG Bull 37:410–432

Fertl WH (1976) Abnormal formation pressures. Elsevier Scientific Publishing Co., New York, p 210

Fulton PM et al (2005) Crustal dehydration and overpressure development on the San Andreas Fault

Gaarenstroom L, Tromp RAJ, De Jong MC, Brandenburg AM (1993) Overpressures in the North Sea: implications for trap integrity. In: Parker JR (ed) Petroleum geology of Northwest distribution and origin of overpressure in the Central Graben of the North Sea Europe: proceedings of the 4th conference. Geological Society of London, pp 1305–1314

Guo XW, He S, Liu KY (2010) Oil generation as the dominant overpressure mechanism in the Cenozoic Dongying depression, Bohai Bay Basin, China. AAPG Bull 94:1859–1881

Hao F, Sun YC, Li ST, Zhang QM (1998) Overpressure retardation of organic-matter maturation and hydrocarbon generation: a case study from the Yinggehai and Qiongdongnan basins, offshore South China Sea. AAPG Bull 79:551–562

Hao F, Li ST, Gong ZS, Yang JM (2000) Thermal regime, inter reservoir compositional heterogeneities, and reservoir filling history of the Dongfang gas field, Yinggehai Basin, South China Sea: evidence for episodic fluid injections in overpressured basins. AAPG Bull 84:607–626

Hennig A, Addis MA, Yassir N, Warrington AH (2002) Pore-pressure estimation in an active thrust region and its impact on exploration and drilling. In: Huffman AR, Bowers GL (eds) Pressure regimes in sedimentary basins and their prediction: AAPG Memoir 76, pp 89–105

Hermanrud C, Wensaas L, Teige GMG, Vik E, Nordgård-Bolås HM, Hansen S (1998) Shale porosities from well logs on Haltenbanken (offshore mid-Norway) show no influence of overpressuring. In: Law BE, Ulmishek GF, Slavin VI (eds) Abnormal pressures in hydrocarbon environments: AAPG Memoir 70, pp 65–85

Herring EA (November 1973) North sea abnormal pressures determined from logs. Pet Eng 45:72–84

Higgins GE, Saunders JB (1967) Report on 1964 Chatham Mud Island, Erin Bay, Trinidad, West Indies. AAPG Bull 51:55–64

Holm GM (1996) The Central Graben—a dynamic overpressure system. http://www.searchanddiscovery.com/abstracts/html/1995/annual/abstracts/0042c.htm

Holm GM (1998) Distribution and origin of overpressure in the Central Graben of the North Sea. In: Law BE, Ulmishek GF, SlavinVI (eds) Abnormal pressures in hydrocarbon environments: AAPG Memoir 70, pp 123–144

Hottman CE, Johnson RK (1965) Estimation of formation pressures from log-derived shale properties. J Pet Technol 17:717–722

Hubbert MK, Rubey WW (1959) Role of fluid pressure in mechanics of overthrust faulting. I. Mechanics of fluid-filled porous solids and its applications to overthrust faulting. GSA Bull 70:167–206

Hunt JM (1990) Generation and migration of petroleum from abnormally pressured fluid compartments. AAPG Bull 74:1–12

John A, Kumar A, Karthikeyan G, Gupta P (2014) An integrated pore pressure model and its application to hydrocarbon exploration: a case study from the Mahanadi Basin, east coast of India, paper appears in Interpretation. 2, Society of Exploration Geophysicists and American Association of Petroleum Geologists, pp SB17–SB26

Kadri IB (1991) Abnormal formation pressures in post-Eocene Formation, Potwar Basin, Pakistan. In: SPE/IADC drilling conference, 21920, pp 213–220

Kong L, Chen H, Ping H, Zhai P, Liu Y, Zhu J (2018) Formation pressure modeling in the Baiyun Sag, Northern South China Sea: implications for petroleum exploration in deep-water areas. Mar Pet Geol 97:154–168

Law BE, Ulmishek GF, Slavin VI (1998) Abnormal pressures in hydrocarbon environments. In: Law BE, Ulmishek GF, SlavinVI (eds) Abnormal pressures in hydrocarbon environments. AAPG Memoir 70, pp 1–11

Leonard RC (1993) Distribution of sub-surface pressure in the Norwegian Central Graben and applications for exploration. In: Parker JR (eds) Petroleum geology of the northwest Europe: proceedings of the 4th conference. Geological Society of London, pp 1295–1303

Liechti P, Roe FW, Hailes NS (1960) Geology of Sarawak, Brunei and western North Borneo. Brit. Borneo Geol. SUTV. Bull 3. https://trove.nla.gov.au/work/22180854?q&versionId=26763739. Accessed 9 Jan 2019

Liu FN (1993) Identify potential sites of gas accumulation in overpressure formation in Qiongdongnan Basin of South China Sea. AAPG Bull 77:888–895

Luo XR, Dong WL, Yang JH, Yang W (2003) Overpressuring mechanisms in the Yinggehai Basin, South China Sea. AAPG Bull 87:629–645

Malick AM (1979) Pressures plague Pakistan's Potwar. In: Petroleum engineer, international, pp 26–36

Mann DM, Mackenzie AS (1990) Prediction of pore pressures in sedimentary basins. Mar Pet Geol 7:55–65

Mattavelli L, Novelli L, Anelli L (1991) Occurrence of hydrocarbons in the Adriatic basin. In: Spencer AM (ed) Generation, accumulation, and production of Europe's hydrocarbons. Oxford University Press, Oxford, pp 369–380

Mukherjee S (2013) Channel flow extrusion model to constrain dynamic viscosity and Prandtl number of the Higher Himalayan Shear Zone. Int J Earth Sci 102:1811–1835

Mukherjee S (2015) A review on out-of-sequence deformation in the Himalaya. In: Mukherjee S, Carosi R, van der Beek P, Mukherjee BK, Robinson D (eds) Tectonics of the Himalaya. Geological Society, London, Special Publications 412, pp 67–109

Mukherjee S, Kumar N (2018) A first-order model for temperature rise for uniform and differential compression of sediments in basins. Int J Earth Sci 107:2999–3004

Nambiar KR, Singh BK, Goswami RN, Singh KRK (2011) Distribution of overpressure and its prediction in Saurashtra Dahanu Block, Western Offshore basin, India. Adapted from extended abstract Presented at

Geo-India, Greater Noida, New Delhi, India, 12–14 Jan 2011

Nashaat M (1998) Abnormally high fluid pressure and seal impacts on hydrocarbon accumulations in the Nile Delta and North Sinai basins, Egypt. In: Law BE, Ulmishek GF, Slavin VI (eds) Abnormal pressures in hydrocarbon environments: AAPG Memoir 70, pp 161–180

Peel FJ, Matthews S (1999) Structural styles of traps in deepwater fold/thrust belts of the northern Gulf of Mexico [abs.]: Extended abstracts Volume. In: AAPG International Conference and Exhibition, Birmingham, p 392

Polutranko AJ (1998) Causes of formation and distribution of abnormally high formation pressure in petroleum basins of Ukraine. In: Law BE, Ulmishek GF, Slavin VI (eds) Abnormal pressures in hydrocarbon environments: AAPG Memoir 70, pp 181–194

Powers MC (1967) Fluid-release mechanism in compacting marine mud-rocks and their importance in oil exploration. AAPG Bull 51:1240–1245

Ramdhan AM, Goulty NR (2010) Overpressure generating mechanisms in the Peciko field, Lower Kutai basin, Indonesia. Pet Geosci 16:367–376

Ramdhan AM, Goulty NR, Hutasoit LM (2011) The challenge of pore pressure prediction in Indonesia's Warm Neogene basins. In: Proceedings of Indonesian petroleum association, 35th annual convention, IPA11-G-141

Rao GN (2001) Sedimentation, stratigraphy, and petroleum potential of Krishna-Godavari basin, East Coast of India. AAPG Bull 85:1623–1643

Rizzi PW (1973) Hochdruckzonen-Früherkennung in Mitteleuropa. Erdol Erdgas Z 89:249–256

Rowan MG, Trudgill BD, Fiduk JC (2000) Deep water salt cored fold belts, northern Gulf of Mexico. In: Mohriak W, Talwani M (eds) Atlantic rifts and continental margins: American Geophysical Union Geophysical Monograph, vol 115, pp 173–191

Roy DK, Ray GK, Biswas AK (2010) Overview of overpressure in Bengal basin, India. J Geol Soc India 75:644–660

Sahay B (1999) Pressure regimes in oil & gas exploration. Allied Publisher Ltd

Sahay B, Fertl WH (1988) Origin and evaluation of formation pressures. Kluwer Academic Publishers, Dordecht, p 292

Saito S, Goldberg D (1997) Evolution of tectonic compaction in the Barbados accretionary prism: estimates from logging-while-drilling. Earth Planet Sci Lett 148:423–432

Schaar G (1977) The occurrence of hydrocarbons in overpressured reservoirs of the Baram Delta (offshore Serawak, Malaysia). In: Proceedings of the annual convention—Indonesian Petroleum Association vol 5, p 1. https://eurekamag.com/research/020/401/0204010 11.php

Singha DK, Chatterjee R (2014) Detection of overpressure zones and a statistical model for pore pressure estimation from well logs in the Krishna-Godavari basin, India. Geochem Geophys Geosyst 15:1009–1020

Summa LL, Pottorf RJ, Schwarzer TF, Harrison WJ (1993) Paleohydrology of the Gulf of Mexico Basin: development of compactional overpressure and timing of hydrocarbon migration relative to cementation. In: Doré AG et al (eds) Basin Modeling: Advances and applications, Norwegian Petroleum Society (NPF) Special Publication 3, pp 641–656

Suppe J, Wittke JH (1977) Abnormal pore fluid pressures in relation to stratigraphy and structure in the active fold and thrust belt of northwestern Taiwan. Pet Geol Taiwan 14:11–24

Swarbrick RE, Osborne MJ (1998) Mechanisms that generate abnormal pressures: and overview. In: Law BE, Ulmishek GF, Slavin VI (eds) Abnormal pressures in hydrocarbon environments: AAPG Memoir 70, pp 13–34

Traugott M (1997) Pore/fracture pressure determinations in deep water. Deepwater Technology, Supplement to August. World Oil 218:68–70

Unruh J, Davisson M, Criss K, Moores E (1992) Implications of perennial saline springs for abnormally high fluid pressures and active thrusting in western California. Geology 20:431–434

Van Ruth PJ, Hillis RR, Swarbrick RE (2002) Detecting overpressure using porosity based techniques in the Carnarvon basin, Australia. APPEA J 42:559

Zhang QM, Liu FN, Yang JH (1996) Overpressure systems and petroleum accumulation in the Yinggehai Basin. Chin Mar Oil Gas (Geol) 10:65–75

Investigation of Erosion Using Compaction Trend Analysis on Sonic Data

6

Abstract

Compaction trends of sediments can decode the mechanism of compaction. Not all kinds of log detect all types of porosity, For example, while Neutron-, sonic- and density logs can decipher porosity, sonic tool cannot detect secondary porosity. Tectonic and isostatic uplift affect petroleum system. The Velocity-depth data from different terrains has been used in studying erosion of petroliferous basins. Porosity-depth trends in well data can indicate the amount of eroded sediment layer. How different authors estimated the thickness of the eroded overburden following different principles is discussed in this chapter.

6.1 Introduction

Processes affecting compaction can be quantified by constructing compaction trends (Magara 1976). Apart from delineating undercompacted zones by comparing the modeled and the actual compaction trends (Magara 1976) and finding out porosity reduction with burial (Issler 1992), constraining the erosion from porosity data would be another approach (Magara 1976). There are few assumptions in this study that the physical properties of the parameters (porosity, sonic velocity, density of bulk rock, etc.) change normally and irreversibly.

Neutron-, sonic- and density logs can decipher porosity. Sonic data is preferred over neutron- and density data since neutron logs in shales are affected by compaction negligibly (Rider 1986), and density logs decipher the total porosity i.e., secondary as well as intergranular (Rider 1986). Acoustic waves emitted by the sonic tool travels through the intergranular porosity and cannot detect secondary porosities like fractures.

Erosion and uplift due to isostasy (Mukherjee 2017) are common geological processes associated with mountain building (Mukherjee 2015) and because of lack of measurable data in many cases, the relationship between uplift and erosion is not quantified strictly. Studies on present day landscapes like the Colorado Plateau (Pederson et al. 2002), Tibetan Plateau (Wobus et al. 2005) and Sierra Nevada (Riebe et al. 2000; Cecil et al. 2006) are good spots for direct measurements. Uplift and erosion also play major roles in oil industry as the magnitude of uplift affects the entire petroleum system [as detailed by Doré and

© Springer Nature Switzerland AG 2020
T. Dasgupta and S. Mukherjee, *Sediment Compaction and Applications in Petroleum Geoscience*,
Advances in Oil and Gas Exploration & Production, https://doi.org/10.1007/978-3-030-13442-6_6

Jensen (1996)]. Velocity-depth being the most available data, it has been used widely for exhumation studies in several petroliferous basins (Hillis 1995a, b; Hansen 1996; Heasler and Kharitonova 1996; Evans et al. 1997; Japsen 1998). This chapter mainly covers exhumation estimation from sonic porosity logs.

6.2 Exhumation Estimation from Porosity Logs

Sediment compaction (e.g., Mukherjee and Kumar 2018) processes are irreversible, and the amount of missing sedimentary section can be delineated from porosity-depth trends in well data. Shales undergo maximum compaction at greater burial depth. In the uplifted and eroded sections there will be a porosity lower than that at deeper depth. Many logs (density, neutron, etc.) can estimate porosity. Sonic transit time is widely used for downhole porosity determination for defining shale compaction trends yielding the rate and the amount of erosion (e.g., Magara 1976; Issler 1992; Nelson and Bird 2005). Sonic transit time depends on factors such as porosity, lithology, fluid content, temperature and overburden pressure. Initially sonic transit time was related to porosity via the Wyllie equation (Wyllie et al. 1956, 1958). Other linear equations also exist (Issler 1992). The acoustic formation factor defined by Raiga-Clemenceau et al. (1988) provides means to calculate porosity (ϕ) from sonic transit time (t) as a function rock type [review by Nelson and Bird (2005)].

$$\phi = 1 - (t_{ma}/t)^{1/x} \qquad (6.1)$$

Here t_{ma} corresponds to the sonic transit time of the rock matrix with zero porosity and where x reflects an exponent specific to the matrix lithology. Issler (1992) adopted this approach to derive shale porosities in the Beaufort-Mackenzie basin, N Canada. Shale compaction trends were

established using these porosities. The main aim of that study was to estimate the amount of erosion (thickness of lost rock column) from the sonic data. A reference compaction curve using sonic porosity data was made from a relatively undeformed and stable part of the basin. This was then compared with the well data from other parts of the well. The vertical offset between the reference well and the well under study represents the amount of erosion. Rowan et al. (2003) later tried to distinguish the lithological effect in the compaction trends. They created the volume of shale curves using gamma ray and derived the matrix transit time and exponent specific to lithology in the un-eroded/still preserved section. Logs were edited in the permafrost and washout zones with abnormal porosities and also in the undercompacted zone with pore fluid overpressuring. Separate compaction trends were established for shales, siltstones, sandstones:

$$\phi_z = \phi_0 \exp(-bz) \qquad (6.2)$$

Here ϕ_z: porosity (as a ratio) at depth z, ϕ_0: porosity at the surface; b: a constant. Other authors (e.g., Falvey and Middleton 1981; Hunt et al. 1998) explained the porosity-depth trends with linear and reciprocal formulas.

6.3 Transit Time and Shale Compaction

Magara (1976) proposed an exponential relationship between transit time and depth for shales

$$\Delta t = \Delta t_0 \exp(-cz) \qquad (6.3)$$

Here Δt: transit time measured by the sonic log, Δt_0: transit time at the present sedimentary surface, c: compaction coefficient, and z: burial depth. Δt_0 is the transit time and the value is near to the transit time of water (Magara 1976) i.e., 189 μs/ft.

Rearranging Eq. (6.3), the maximum burial depth Z_{max} can be calculated as

$$Z_{max} = \ln(\Delta t/\Delta t_{0N})/(-c) \quad (6.4)$$

And the net uplift and erosion (Δz) are calculated by subtracting the present burial depth (z)

$$\Delta z = \ln(\Delta t/\Delta t_{0N})/(-c)-z \quad (6.5)$$

Here Δt_{0N}: Δt_0 magnitude for normally compacted shales.

Later the Eq. (6.3) above was modified as:

$$\Delta t = \Delta t_0 \exp(-cz)+c \quad (6.6)$$

Magara (1976) stated matrix transit time ranging between 39 and 68 μs/ft for dolomites and shales. Other parameters being the same as Eq. (6.6), c is shift constant nearly equal to the rock matrix transit time.

6.4 Methodology

Erosion calculation technique and its validation: Magara (1976) proposed a method for erosion estimation using sonic logs. Figure 6.1a depicts the situation where the sediments are normally compacted, and the surface transit time is

denoted by t_0. Magara (1976) stated the value of t_0 to be ~189 μs/ft. Figure 6.1b represents a situation where the upper surface has been eroded and the thickness of the eroded part can be estimated from the vertical distance between the eroded surface and the original depositional surface.

Heasler and Kharitonova (1996) studied 36 wells in the Bighorn Basin, Wyomingto estimate the erosional thickness. The normal compaction trend was established using Eq. 6.6. The value of the surface transit time t_0 ranges ~180 to 210 μs ft^{-1} and it is equivalent to the transit time of pure water (Magara 1976). Heasler and Kharitonova (1996) estimated the erosional thickness by fitting curves to the data. Equation (6.3) with no shift constant and Eq. (6.6) with a shift constant of 64 μs ft^{-1} showed erosional thickness of ~6400 to 3400 ft, respectively in Emblem State 1. In the Bridger Butte Unit 3 well, Eq. (6.3) with zero shift constant and Eq. (6.6) resulted in erosional thickness of 5000 and 1200 ft, respectively (Fig. 6.2).

Analysis of 36 wells using Eq. (6.6) showed an erosional thickness ranging between 389 and 3600 m. whereas using Eq. (6.3) shows 0–1200 m erosional thickness. The erosional estimates from this methodology was contoured and the Bighorn Basin showed 900–1200 m of erosion in the N/NW part. The basin center and the W

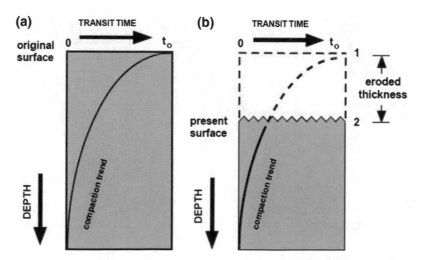

Fig. 6.1 Erosion thickness calculation from compaction trend curve derived from sonic transit time (Magara 1976). **a** No erosion and the surface transit time is equal to that of water. **b** Upper surface is eroded and the thickness is the vertical distance between the present surface and the original surface

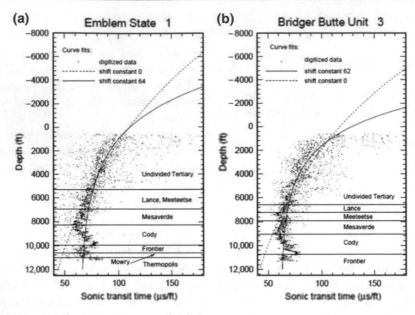

Fig. 6.2 Compaction trend and erosional thickness from **a** Emblem State1 and **b** Bridger Butte Unit 3 by Heasler and Kharitonova (1996)

margin showed 0–900 m of erosion. These estimates match geomorphologic studies. Smith and Braile's (1993) study also states that the NW part of the basin is maximum eroded and the erosion estimation from different studies like from hydrocarbon maturation studies in Bighorn Basin (Hagen and Surdam 1984; Hagen 1986) also supported by Heasler and Kharitonova (1996).

The main problem faced by the later authors was that the study area mostly consisted of mixed lithologies and later the authors found that the erosion thickness estimation using shale data in few wells resemble results from mixed lithologies.

Burns et al. (2005) developed a method for erosional thickness estimation of the Upper Cretaceous and Tertiary lithology of the Colville

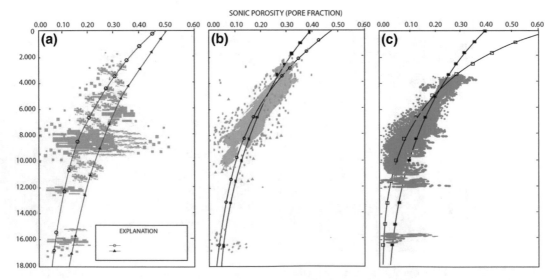

Fig. 6.3 Porosity versus depth curve by Burns et al. (2005) for different lithology using volume of shale cut off. **a** Sand ($v_{shale} < 0.01$), **b** Siltstone ($0.49 \ v_{shale} < 0.51$) and **c** shale ($v_{shale} > 0.51$) along with the curves by Rowan et al. (2003)

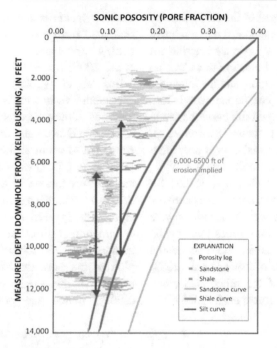

Fig. 6.4 Visual comparison of sonic porosity data and Rowan et al. (2003) curves for each lithology. The vertical offset is the erosional thickness implied

Basin of northern Alaska using the porosity depth trends derived from sonic logs. Rowan et al. (2003) derived the normal compaction trend from sonic porosity depth trends in wells drilled in a minimally/non-exhumed area. The normal compaction trend from non-exhumed part was

Fig. 6.5 Sonic velocity data and porosity from sonic data. The porosity trend can be approximated as linear trend (Li et al. 2007)

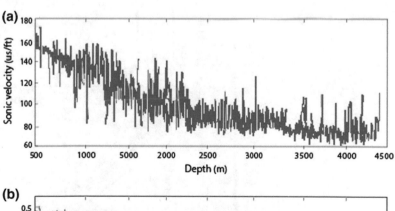

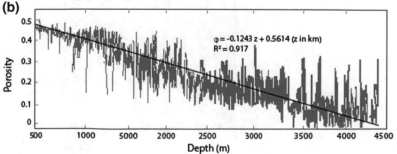

fitted to the wells in the rest of the study area by adjusting the datum of the modeled curve relative to the ground surface. The main assumption for this work was that after exhumation there was no significant sediment deposition, and the thickness of the eroded sediments can be estimated from the vertical distance between the current ground surface and the adjusted datum. Rowan et al. (2003) defined the exponential curves according to lithology using the volume of shale i.e., v_{shale} $_{cutoff}$ < 0.01, > 0.99 and 0.49 < v_{shale} < 0.51 for sands, shales and siltstones, respectively (Fig. 6.3). Data from 145 wellsite locations were studied where the anomalous zones and invalid porosities were edited. Later, the exponential normal compaction trends of Rowan et al. (2003) were plotted for each of the lithologies along with the porosity data by Burns et al. (2005). The

vertical thickness representing the erosion was estimated from visual comparison of the Rowan et al.'s curves and the porosity data from Burns et al. (2005) for each lithology group (Fig. 6.4). Burns et al. (2005) could not verify the exhumation estimation in most of the places from this methodology because of the absence of true reference values. Later borehole stratigraphy was examined to estimate the erosional thickness.

Li et al. (2007) estimated the erosional thickness from sediments of Xihu depression, E China Sea Basin, which is a tectonically inverted zone. Li et al. (2007) used a similar methodology for exhumation estimation as used by previous authors using the sonic transit time. Rather than plotting the porosity-depth data as exponential curves, which is the most common way, Li et al. (2007) plotted the porosity depth data for

Fig. 6.6 Linear porosity trend of 26 wells from Xihu depression area (Li et al. 2007)

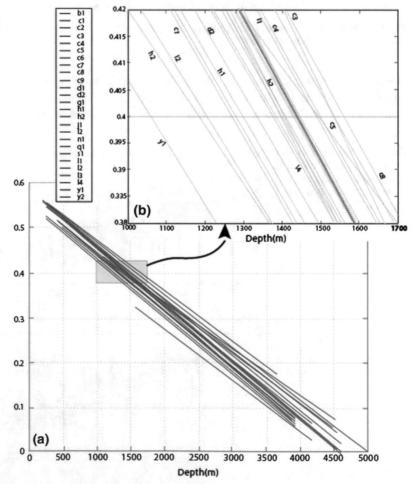

siliciclastic rocks from 26 wells of the Xihu depression area with linear fitting (Figs. 6.5 and 6.6) because linear models give better results. The linear porosity-depth curve showed a similar porosity gradient since these curved are almost parallel. But the offset between each of the linear curves reveals differential uplift within tectonically inverted zones. Availability of dense multi-channel seismic data provides a direct measurement and validation of erosion estimation methodology. Li et al. (2007) concluded that the erosion is larger at the northern part of the central inverted zone of the Xihu depression and the results are comparable with other studies in the Xihu depression. The erosional thickness ranges from 0 to 501.3 m within the 26 wells studied.

References

Burns WM, Hayba DO, Rowan EL, Houseknecht DW (2005) Studies by the U.S. Geological Survey in Alaska: U.S. Geological Survey Professional Paper 1732–D: Estimating the Amount of Eroded Section in a Partially Exhumed Basin from Geophysical Well Logs: An Example from the North Slope

Cecil MR, Mihai ND, Peter WR, Clement GC (2006) Cenozoic exhumation of the northern Sierra Nevada, California, from (U-Th)/He thermochronology. GSA Bull 118:1481–1488

Doré AG, Jensen LN (1996) The impact of late Cenozoic uplift and erosion on hydrocarbon exploration. Global Planet Change 12:415–436

Evans D, Morton AC, Wilson S, Jolley D, Barreiro BA (1997) Palaeoenvironmental significance of marine and terrestrial Tertiary sediments on the NW Scottish Shelf in BGS borehole 77/7. Scott J Geol 33:31–42

Falvey DA, Middleton MF (1981) Passive continental margins; evidence for a prebreakup deep crustal metamorphic subsidence mechanism. Oceanol Acta 4:103–114

Hagen ES (1986) Hydrocarbon maturation in Laramide-style basins—Constraints from the northern Bighorn Basin, Wyoming and Montana: Laramie, University of Wyoming, Ph.D. thesis, 215 p

Hagen ES, Surdam RC (1984) Maturation history and thermal evolution of cretaceous source rocks of the Bighorn Basin, Wyoming and Montana. In: Woodward J, Meissner FF, Clayton JL (eds) Hydrocarbon source rocks of the greater Rocky Mountain region. Rocky Mountain Association of Geologists, pp 321–338

Hansen S (1996) Quantification of net uplift and erosion on the Norwegian Shelf south of 66°N from sonic transit times of shale: Norsk Geologisk Tidsskrift 76:245–252

Heasler HP, Kharitonova NA (1996) Analysis of sonic well logs applied to erosion estimates in the Bighorn Basin, Wyoming. AAPG Bull 80:630–646

Hillis RR (1995a) Quantification of tertiary exhumation in the United Kingdom southern North Sea using sonic velocity data. AAPG Bull 79:130–152

Hillis RR (1995b) Regional Tertiary exhumation in and around the United Kingdom. In: Buchanan JG, Buchanan PG (eds) Basin inversion, vol 88. Geological Society Special Publication, London, pp 167–190

Hunt JM, Whelan JK, Eglinton LB, Cathles LM III (1998) Relation of shale porosities, gas generation, and compaction to deep overpressures in the U.S. Gulf Coast. In: Law BE, Ulmishek GF, Slavin VI (eds) Abnormal pressures in hydrocarbon environments, vol 70. American Association of Petroleum Geologists Memoir, pp 87–104

Issler DR (1992) A new approach to shale compaction and stratigraphic restoration, Beaufort-Mackenzie Basin and Mackenzie Corridor, Northern Canada. Am Asso Petrol Geol Bull 76:1170–1189

Japsen P (1998) Regional velocity-depth anomalies, North Sea Chalk: a record of overpressure and Neogene uplift and erosion. AAPG Bull 82:2031–2074

Li CF, Zhou Z, Ge H, Mao Y (2007) Correlations between erosions and relative uplifts from the central inversion zone of the Xihu Depression, East China Sea Basin. Terre Atmos Oceanic Sci 18:757–776

Magara K (1976) Thickness of removed sedimentary rocks, paleopore pressure, and paleotemperature, southwestern part of Western Canada basin. Am Assoc Pet Geol Bull 60:554–565

Mukherjee S (2015) Petroleum geosciences: Indian contexts. Springer Geology. ISBN 978-3-319-03119-4

Mukherjee S (2017) Airy's isostatic model: a proposal for a realistic case. Arab J Geosci 10:268

Mukherjee S, Kumar N (2018) A first-order model for temperature rise for uniform and differential compression of sediments in basins. Int J Earth Sci 107:2999–3004

Nelson PH, Bird KJ (2005) Porosity-depth trends and regional uplift calculated from sonic logs, national petroleum reserve in Alaska, Scientific Investigations Report 20055051, US Geological Survey

Pederson JL, Mackley RD, Eddleman JL (2002) Colorado Plateau uplift and erosion evaluated using GIS. GSA Today 12:4–10

Raiga-Clemenceau J, Martin JP, Nicoletis S (1988) The concept of acoustic formation factor for more accurate porosity determination from sonic transit time data. In: SPWLA 27th annual logging symposium. Society of Petrophysicists and Well-Log Analysts, vol 29, pp 54–60

Rider MH (1986) The geological interpretation of well logs. Blackie & Son Limited, London, p 175

Riebe CS, Kirchner JW, Granger DE, Finkel RC (2000) Erosional equilibrium and disequilibrium in the Sierra Nevada, inferred from cosmogenic 26Al and 10Be in alluvial sediments. Geology 28:803–806

Rowan EL, Hayba DO, Nelson PH, Burns WM, Houseknecht DW (2003) Sandstone and shale compaction curves derived from sonic and gamma ray logs in offshore wells, North Slope, Alaska—parameters for basin modeling: U.S. Geological Survey Open-File Report 03–329

Smith RB, Braile LW (1993) Topographic signature, space-time evolution, and physical properties of the Yellowstone-Snake River Plain volcanic system:

the Yellowstone hotspot. In: Snoke AW, Steidtmann J, Roberts SM (eds) Geology of Wyoming: Geological Survey of Wyoming Memoir No. 5, pp 694–754

Wobus C, Heimsath A, Whipple K, Hodges K (2005) Active out-of-sequence thrust faulting in the central Nepalese Himalaya. Nature 434:1008–1011

Wyllie MRJ, Gregory AR, Gardner LW (1956) Elastic wave velocities in heterogeneous and porous media. Geophysics 21:41–70

Wyllie MRJ, Gregory AR, Gardner GHF (1958) An experimental investigation of factors affecting elastic wave velocities in porous media. Geophysics 23:459–493

Abstract

Before 1980s geoscientists believed the main reason of overpressure was drilling-induced. Subsequently other processes, e.g., compaction disequilibrium and aquathermal expansion, responsible for overpressure were established. This chapter reports overpressure condition from two kinds of plate margins from several places in the world: (i) collisional margins especially at the accretionary prisms, subduction zones and decollement zones; and (ii) extensional margins such as rifts and passive margins where growth faults boundaries of continental shields are the zones of overpressure.

7.1 Introduction

Concepts on the genesis and development of abnormal pressure have evolved through time. The main cause for development of overpressure in literature before 1980s was mainly because of drilling and completion-related procedures. As discussed in Chaps. 3 and 4, there are several reasons for overpressure generation: compaction disequilibrium and processes involving aquathermal expansion etc. During recent studies it is evident that the total processes involved in the petroleum system analysis starting from hydrocarbon generation from source rocks, expulsion and migration and ultimately the entrapment of hydrocarbon also can create abnormal pressure (Osborne and Swarbrick 1997). Studies about abnormal pressure are important in hydrocarbon exploration (Law et al. 1998).

Abnormally pressured reservoirs are distributed worldwide in several geological conditions as well as within rocks of diverse ages (Law et al. 1998). A spatial disparity in compaction rate in sedimentary basins is a common phenomenon (e.g., Mukherjee and Kumar 2018). Law et al. (1998) mentioned abnormally pressured environments in deltaic systems where compaction disequilibrium is one of the dominant mechanisms for overpressure. They also referred that in the U.S. Gulf coast hydrocarbon generation leads to overpressure. Regions where basin centered gas accumulation takes place are Alberta basin of Canada, Greater Green River Basin of Wyoming, Colorado and these zones are mainly concentrated in North America. Abnormal pressures are also evident in unconventional reservoirs such as coal bed methane, shale gas as seen in the Appalachian Basin, where oil and gas is produced from organically rich Devonian shale (Reeves et al. 1996). The gas productive Barnett shale is another example of abnormally pressured shale.

T. Dasgupta and S. Mukherjee, *Sediment Compaction and Applications in Petroleum Geoscience*,
Advances in Oil and Gas Exploration & Production, https://doi.org/10.1007/978-3-030-13442-6_7

7.2   Plate Margins

Overpressure conditions have been reported also from different plate margins.

7.2.1   Convergent Margins

Overpressured conditions along convergent plate margins arise mainly due to active compressional stress regime. Other reasons are due to shale and salt diapirism (Mukherjee et al. 2010; Mukherjee 2011), massive dumping of sediments, especially in adjacent foreland basins, change in fluid density and also due to facies change or variation in clay mineralogy.

In the E coast basin (New Zealand), it is a prolific petroleum system but it is non-commercial because of the complexities in structural and stratigraphy context (Burgreen-Chan et al. 2016) and near the Hikurangi subduction zone, the exploration activities are hindered due to occurrence of overpressure zones (Darby and Beavan 2001). A high mud weight, as high as 19.5 ppg, was required to control the influx of formation fluids into the wells. In this basin, the occurrence of overpressure is totally controlled by lithostratigraphic distribution and low permeability mudstones. The major pore pressure generating mechanism in this area i.e. compaction disequilibrium is due to the Neogene evolution of the plate margin where the high and fast sedimentation happened during the Miocene. The extreme overpressure condition is due to the uplift of under-compacted sediments up to 3000 m by Neogene compression at $\sim$ 1000 m Ma^{-1}. Some imprints of lateral tectonic compression also caused high overpressure by shearing of thick mudstones (Darby and Beavan 2001).

Accretionary prisms are another zone of overpressure but reliable measurements from the youngest as well the active part of the accretionary prisms are lacking (Moore and Vrolijk 1992). The under-thrusting of Upper Cretaceous Atlantic zone beneath the Caribbean plate produced the Barbados accretionary prism (Larue and Speed 1984). This accretionary prism mostly consists of Quaternary-Upper Miocene

calcareous mudstone and the detachment between the accretionary prism and the decollement zone mainly occurs at $\sim$ 500 to 530 m depth (Moore et al. 1995). In the site 671, faults are predominant in the zones of high overpressured areas (Fig. 7.1).

The general morphology and the mechanics working at the subduction zone are mainly controlled by pore fluid pressure (Davis et al. 1983). The location of the main decollement zone is mainly above the zone of high fluid content. Several subduction zones with elevated pore pressure conditions have been reported. Screaton et al. (2002) mentioned excess pore pressure from the Nankai margin where the porosity-depth profile was drawn to estimate the excess pore pressure conditions in the under-thrust zones. Similar attempt was made by Cochrane et al. (1996) where the compressional

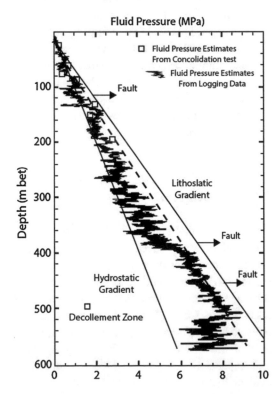

Fig. 7.1 The occurrence of high overpressured zones is corresponding to the fault zones. The fluid pressure estimates from consolidation tests are derived from analysis of individual samples and is in agreement with the log derived fluid pressures (Moore et al. 1995)

wave velocity was inverted to porosity at the Cascadia subduction zone (stretching from northern Vancouver Island to northern California) and porosity data was plotted against depth. The logging while density (LWD) density tool gave density and that was converted into porosity by Moore and Tobin (1997) at the Barbados accretionary prism and these porosities were used for pore pressure studies. Saffer (2003) discussed changes in the physical properties from the Costa Rica, Nankai and Barbados subduction zones and merged the lab results with the LWD data. Ocean drilling program data (site 1043) were included. According to Saffer (2003), lab analyses and log signatures indicate that the margin wedge drained heterogeneously. There is differential drainage pattern in the upper and the lower parts. These studies are supported by the effective stress measurements from laboratories. The decollement zone occurs at the plate boundary and is the pathway for fluid migration and the fluid pressures are generally high in this region and also occur in spikes at 505 and 515 m (Fig. 7.2).

The calculated fluid pressure was as high as > 90% of lithostatic pressure below the thrust in the prism. There are many thin intervals which show anomalously low density and resistivity in the logs.

Another place with high pore pressure is the region where the Philippine sea plate is subducting beneath Japan as it forms the Nankai Trough and the accretionary prism is also associated with overpressure (Moore et al. 2012). Several scientists quoted that the major slip of the 1944 earthquake with 8.1 magnitude occurred through the mega-splay fault. Another site C0001 is also located above the mega-splay. There are several low resistivity zones with low sonic velocity. The deeper sections of the site C0001 and C0003 show the divergence from the standard pressure gradient (Moore et al. 2012). Huge pressure anomalies are seen in holes C0001A and C0001D (Fig. 7.3). Both the holes maintain a standard pressure gradient up to 400 mbsf and the level of pressure fluctuations are higher at places when it crosses the lithostatic pressure (Fig. 7.4). Seismic lines through the sites C0001A and C0001D suspect of fault/fracture zones/lineaments. These could be the potential sites for fluid migration and thus causing overpressure.

Pore pressures are elevated near the subduction zone in offshore Indonesia, E Taiwan, S America and the Philippine islands (Fertl et al. 1976). Besides the Japan trench, other DSDP sites were not logged. As mentioned above the Japan trench shows decrease in density along with overpressured fluids present in fractured mudstones (Carson et al. 1982). Overpressured fluids were estimated from the pore fluids

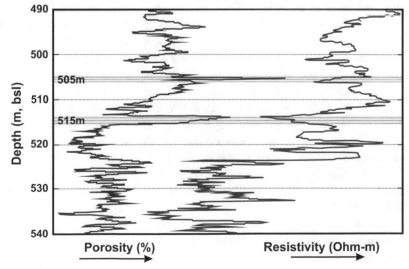

Fig. 7.2 Spike at 505 and 515 m which shows increase in porosity and decrease in resistivity through decollement zone (Saffer 2003)

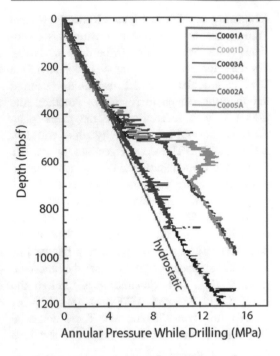

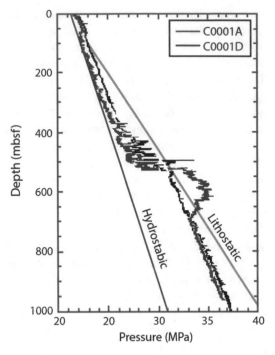

Fig. 7.3 Annular pressure while drilling for the sites. The sites C0002, C0004, C0006 and the shallower part of the C0001 and C0003 follows the standard gradient. There is divergence in the annular pressure while drilling data in the deeper section of C0001 and C0003 section (Moore et al. 2012)

Fig. 7.4 Annular pressure while drilling for the sites C0001A and C0001D. In the site C0001A the standard APWD is there up to 413 mbsf, beyond this depth up to 530 mbsf there is shift in the pressure gradient and at places crosses the lithostatic gradient and later follows the stable pressure gradient. In the hole C0001D the APWD pressure is stable up to 400 mbsf but is followed by fluctuations in pressures while drilling downhole and shoots above lithostatic above 522 mbsf and remains above till 690 mbsf and similar to hole C0001A the pressure falls back to normal gradient (Moore et al. 2012)

obtained during the drilling of Leg 84, firstly by the backpressure measurement when the cuttings are constricted in the drill holes and secondly from water samples and heat flow measurements.

High pore fluid pressures were encountered during drilling of wells in British Columbia and in the Gulf of Alaska. Wells were drilled in varied places starting from continental shelf, over the thrust fault planes. Overpressured zones have also been reported from N Plains of Gorgan, central Iran as well as from the Neocomian Fahliyan Formation of Abadan Plain Basin, SW Iran (Fertl et al. 1976; Soleimani et al. 2017) and elsewhere. Similar overpressure conditions are also reported from Tertiary sediments in Himalayan foothills from N Pakistan and India, from foreland Alpine basins of Germany, Austria and Italy (Fertl et al. 1976).

The activity in Costa Rica convergent margin is dynamic because of the high convergence rate and subduction of the sediments. Numerical

modelling was performed to see the reservoir parameters such as porosities, fluid pressures and the evolution of permeabilities during subduction (Screaton et al. 2002). The permeability-porosity relationship established from the laboratory experiments can be correlated with the pore pressure results from compaction of sediments as well as tests (Screaton et al. 2002). The complete subduction of sediments and the highly porous and permeable sediments are directly below the decollement zone. High rate of fluid flow is there towards the top and along the decollement.

The Northern island of New Zealand is an active convergent margin and the smectite-rich seal creates overpressure by compaction disequilibrium mechanism in the underlying Cretaceous to Paleocene section. Evolution of porosity and overpressure was studied in this basin and a framework was made to understand the reasoning of overpressure generation-related mechanisms. To comprehend the impact of compression on pore pressure and the stresses, poroelastic modelling was attempted. Through geological time the shortening has impacted the relative principal stresses. The changes in the stress regimes characteristically vary the deformational styles and this can modify the tectonic evolution.

7.2.2   Extensional Margins

In extensional margins—both passive margins and rifted settings—abnormal formation pressure result from increase in sediment load, rapid sedimentation, dewatering process during diagenesis and due to lateral and vertical facies variation.

The N/NW shelf of Gulf of Mexico is categorized as an extensional margin (Chilingarian and Wolf 1988). The northern margin showed rapid progradation and consists dominantly of permeable sands. Rapid burial of these sequences along with the presence of low permeable facies increases pressure of the trapped fluid to attain an abnormal magnitude. Mudstones exhibits overpressure where the pressure gradient is ~ 22.5 kPa m^{-1} (Fig. 7.5). Numerous growth faults in this area can either enhance or reduce permeability. The timing of fluid transport along the structural discontinuities plays an important role in vertical fluid transport. As discussed earlier, the most common theories for the overpressure increase in Gulf coast are (i) compaction disequilibrium; (ii) aquathermal pressuring; (iii) mineral-phase transformations; and (iv) hydrocarbon maturation.

In Indian offshore, abnormal pore pressure conditions occur in varied geological contexts. The offshore Krishna–Godavari Basin along the east coast of India consists of thick sediments with high sedimentation rate. The Mio-Pliocene and Cretaceous shales show overpressure zones and these abnormal pressures have been

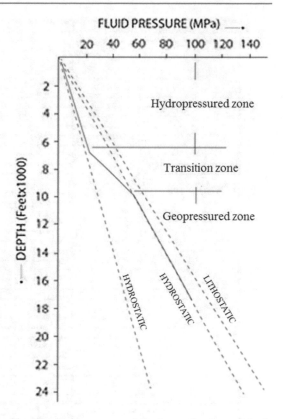

Fig. 7.5 Fluid pressure gradient as a function of depth (Chilingarian and Wolf 1988)

developed by high rate of sediment transport and deposition (also see Fig. 7.14, Bastia and Radhakrishna 2012).

Also a case study by Chatterjee et al. (2012) discusses the high temperature-high pressure (HTHP) regime in Krishna–Godavari Basin and two wells A and B (Figs. 7.6 and 7.7) delineates the overpressure zones from specific stratigraphic units. While the predrill high pore pressure for the drilled well was predicted from the pre-stack time migrated (PSTM) velocity data, the post drill pore pressure plot was generated from logs (resistivity, density, sonic and VSP). Both the pre-drill as well as the post-drill pore pressure curves show overpressure zones. In the well B the overpressure zones are speculated by high fluid pressures in modular dynamic tester (MDT) points as well as kicks encountered during drilling (Fig. 7.7).

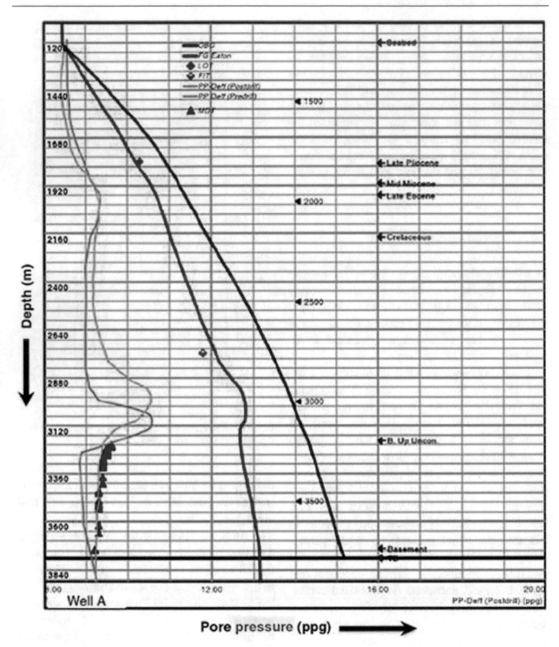

Fig. 7.6 Pore pressure plot of KG basin offshore with the example of well A (Chatterjee et al. 2012)

Other than passive margins, rift settings within and at the periphery of continental cratons are also key areas for development of overpressure. Rift basins comprise of episodic deposition of sedimentary sequences (Dasgupta and Mukherjee 2017) and at times the sedimentation rates may be very high. Moreover isolated/ perched sandstone bodies also exist (Kothari et al. 2016) and heterogeneous clastic reservoirs in rift basins; all these can lead to overpressure conditions. This has been reported from a number of rift basins worldwide, viz., North Sea rift, Red Sea rift, Malay Basin in offshore Malaysia, SE part of Godavari rift in S India, Cambay and

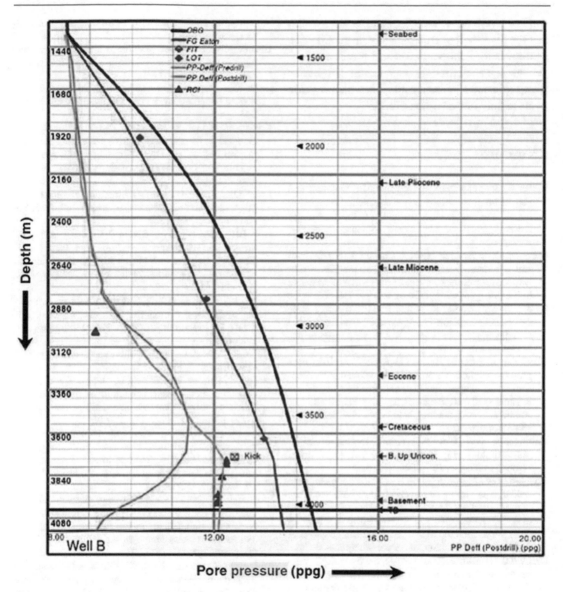

Fig. 7.7 Pore pressure plot of KG basin offshore with the example of well B (Chatterjee et al. 2012)

Barmer Basin from W India (Fertl et al. 1976; Sahay and Fertl 1988).

Petroleum exploration was a major boom in rift basins as many data were available from the North Sea (Glennie 1986). The most common occurrence of abnormal pressure was from both pre-unconformity and post-unconformity sequences. The most common examples of mudrocks are in the Viking graben and this rests above the unconformities of Late Jurassic age as well as of Late Jurassic to Early Cretaceous. As mentioned, the main reservoir of Ekofisk is bound by mudrocks from top and bottom and this makes the entire reservoir highly overpressured. Kimmeridge clay Formation is highly overpressured and radioactive, which is also an important source rock. Among the other rift basins, abnormal pressure is present in offshore fields of the Atlantic shelf of North America (Grant and Madsen 1986) and the NW shelf of Australia (Nyein et al. 1977). Moreover far field compressional forces or changes in stress regime in continental rift basins (Dasgupta and Mukherjee 2017) can also develop overpressure or underpressure.

References

Bastia R, Radhakrishna M (2012) Basin evolution and petroleum prospectivity of the continental prospectivity of the continental margins of India, vol 59. Developments of Petroleum Science, Elsevier, Amsterdam

Burgreen-Chan B, Meisling KE, Graham S (2016) Basin and petroleum system modelling of the East Coast Basin, New Zealand: a test of overpressure scenarios in a convergent margin. Basin Res 28:536–567

Carson B, von Huene R, Arthur M (1982) Small scale deformation structures and physical properties related to convergence in Japan Trench slope sediments. Tectonics I:277–302

Chatterjee A, Mondal S, Basu P, Patel BK (2012) Pore pressure prediction using seismic velocities for deepwater high temperature-high pressure well in offshore Krishna Godavari Basin, India. In: SPE oil and gas conference and exhibition held in Mumbai, India, SPE 153764, pp 1–20

Chilingarian GV, Wolf KH (1988) Developments of sedimentology. In: Diagenesis II, vol 43. Elsevier Science Publishers B.V., Amsterdam

Cochrane GR, Moore JC, Lee HJ (1996) Sediment pore fluid overpressuring and its effect on deformation at toe of the Cascadia accretionary prism from seismic velocities. In: Bebout GE, Scholl DW, Kirby SH, Platt JP (eds) Subduction top to bottom: American geophysical union geophysical monograph, vol 96, pp 57–64

Darby DJ, Beavan J (2001) Evidence from GPS measurements for contemporary plate coupling on the southern Hikurangi subduction thrust and partitioning of strain in the upper plate. J Geophys Res 106:30881–30891

Dasgupta S, Mukherjee S (2017) Brittle shear tectonics in a narrow continental rift: asymmetric non-volcanic Barmer basin (Rajasthan, India). J Geol 125:561–591

Davis D, Suppe J, Dahlen FA (1983) Mechanics of fold and thust belts and accretionary wedges. J Geophys Res 88:1153–1172

Fertl WH, Chilingarian GV, Rieke HH (1976) Abnormal formation pressures implications to exploration, drilling, and production of oil and gas resources, vol 2 (Chap. 9). Developments in Petroleum Science, Elsevier, pp 325–349

Glennie KW (1986) Introduction to the petroleum geology of the North sea. Blackwell Scientific Publications, Hoboken, pp 63–86

Grant WD, Madsen OS (1986) The continental shelf bottom boundary layer. Ann Rev Fluid Mech 18:265–305

Kothari V, Konar S, Naidu B, Desai A, Sunder VR, Goodlad S, Mohapatra P (2016) Lacustrine turbidites in rift basins: genesis, morphlogy and petroleum potential—a case study from Barmer Basin, Conference material

Larue DK, Speed RC (1984) Structure of accretionary prism complex of Barbados, II; Bissex hill. Geol Soc Am Bull 95:1360–1372

Law BE, Ulmishek GF, Slavin VI (1998) Abnormal pressures in hydrocarbon environments. In: Law BE, Ulmishek GF, Slavin VI (eds) Abnormal pressures in hydrocarbon environments. AAPG Memoir 70, pp 1–11

Moore JC, Tobin H (1997) Estimated fluid pressures of the Barbados accretionary prism and adjacent sediments. In: Proceedings of the ocean drilling program, Scientific results, 156, pp 229–238

Moore JC, Vrolijk P (1992) Fluids in accretionary prisms. Rev Geophys 30:113–135

Moore JC, Shipley T, Goldberg D, Ogawa Y et al (1995) Abnormal fluid pressures and fault-zone dilation in the Barbados accretionary prism: evidence from logging while drilling. Geology 23:605–608

Moore CJ, Barrett M, Thu MK (2012) High fluid pressures and high fluid flow rates in the Megasplay Fault Zone, NanTroSEIZE Kumano Transect, SW Japan. Geochem Geophy Geosyst, 13. https://doi.org/10.1029/2012GC004181

Mukherjee S (2011) Estimating the viscosity of rock bodies—a comparison between the Hormuz- and the Namakdan Salt Domes in the Persian Gulf, and the Tso Morari Gneiss Dome in the Himalaya. J Indian Geophys Union 15:161–170

Mukherjee S, Kumar N (2018) A first-order model for temperature rise for uniform and differential compression of sediments in basins. Int J Earth Sci 107:2999–3004

Mukherjee S, Talbot CJ, Koyi HA (2010) Viscosity estimates of salt in the Hormuz and Namakdan salt diapirs, Persian Gulf. Geol Mag 147:497–507

Nyein RK, MacLean L, Warris BJ (1977) Occurrence, prediction and control of geopressures on the Northwest shelf of Australia. APPEA J 17:64–77

Osborne MJ, Swarbrick RE (1997) Mechanisms for generating overpressure in sedimentary basins: a reevaluation. Am Assoc Pet Geol 81:1023–1041

Reeves SR, Kuustraa VA, Hill DG (1996) New basins invigorate U.S gas shales play. Oil Gas J 94:53–58

Saffer DM (2003) Pore pressure development and progressive dewatering in underthrust sediments at the Costa Rican subduction margin: comparison with Northern Barbados and Nankai. J Geophys Res 108:B5

Sahay B, Fertl WH (1988) Origin and evaluation of formation pressures. Kluwer Academic Publishers, Dordecht, Boston, London, p 292

Screaton E, Saffer D, Henry P, Hunze S (2002) Porosity loss within underthrust sediments of the Nankai accretionary complex: implications for overpressures. Geology 30:19–22

Soleimani B, Hassani-Giv M, Abdollahifard I (2017) Formation pore pressure variation of the Neocomian sedimentary succession (the Fahliyan Formation) in the Abadan Plain basin, SW of Iran. Geofluids. https://doi.org/10.1155/2017/6265341

Correction to: Detection of Abnormal Pressures from Well Logs

Correction to:
Chapter 4 in: T. Dasgupta and S. Mukherjee,
Sediment Compaction and Applications in Petroleum Geoscience,
Advances in Oil and Gas Exploration & Production,
https://doi.org/10.1007/978-3-030-13442-6_4

In the original version of the book, Figure 4.10 of Chapter 4 was inadvertently published without proper permission. The Figure has now been replaced with the correct figure after getting permission, and the citations have been added to the reference list. The book and the chapter have been updated with the changes.

The updated version of this chapter can be found at
https://doi.org/10.1007/978-3-030-13442-6_4

© Springer Nature Switzerland AG 2020
T. Dasgupta and S. Mukherjee, *Sediment Compaction and Applications in Petroleum Geoscience,*
Advances in Oil and Gas Exploration & Production, https://doi.org/10.1007/978-3-030-13442-6_8